Current Flow

The Electrification of Palestine

Ronen Shamir

Stanford University Press
Stanford, California

Stanford University Press
Stanford, California

Printed in the United States of America on acid-free, archival-quality paper

Library of Congress Cataloging-in-Publication Data

Shamir, Ronen, author.
Current flow : the electrification of Palestine / Ronen Shamir.
pages cm
Includes bibliographical references and index.
ISBN 978-0-8047-8706-2 (cloth : alk. paper)
1. Electrification—Palestine—History—20th century. 2. Electrification—Social aspects—Palestine. 3. Jewish-Arab relations. 4. Palestine—Social conditions—20th century. I. Title.
HD9685.P192S53 2013
333.793'2095694—dc23

2013026237

Typeset by Newgen in 11/13.5 Adobe Garamond

For Leigh and Omer

"The sun never sets on the British Empire," says Miss Lumley, tapping the roll-down map with her long wooden pointer. In countries that are not the British Empire, they cut out children's tongues, especially those of boys. Before the British Empire there were no railroads or postal services in India. And Africa was full of tribal warfare, with spears, and had no proper clothing. The Indians in Canada did not have the wheel or telephones, and ate the heart of their enemies in the heathenish belief that it would give them courage. The British Empire changed all that. It brought in electric lights.

—*Margaret Atwood,* Cat's Eye

Contents

Figures xi
Acknowledgments xiii

Introduction 1

1 The Jaffa Powerhouse 24

2 Let the Poles Act 47

3 Wired 78

4 From State to Market 104

5 Assembling a Jewish Economy 126

Conclusion 149

Notes 157
References 185
Index 195

Figures

1 Pinhas Rutenberg and staff at the opening ceremony of the Jaffa Electric Company powerhouse, June 10, 1923 25
2 Comparison of English and German machines and tenders 41
3 Letter from Pinhas Rutenberg to Winston Churchill, December 21, 1921 44
4 New generators being transported to the powerhouse 48
5 Settlement agreement between Abdel Magid Abou, a landowner, and the Jaffa Electric Company, March 15, 1925 53
6 Provisional street lighting in Jaffa 57
7 Letter from A. El-Said, mayor of Jaffa, to the Jaffa Electric Company, March 4, 1925 59
8 Comparison of vehicular traffic between Jaffa and Tel Aviv 66
9 Stringing new wires 79
10 Power line extension into Tel Aviv 87
11 Report of directors, Palestine Electric Corporation, for the year ended December 31, 1933 93
12 Excerpt of the electricity ordinance, 1927 95
13 Palestine Railway, 1922 105
14 Report and balance sheet, Palestine Electric Corporation, for the year ended December 31, 1933 129
15 Report and balance sheet, Palestine Electric Corporation, for the year ended December 31, 1935 131

Acknowledgments

I WROTE THIS BOOK ON THE ROAD. The raw data of three years of archival research were backed up on two laptops and two USBs. Such materials are compact to pack and carry nowadays, and the cost of their overweight is borne only in one's mind.

I owe many thanks to the people, academic departments, research institutions, and archives that provided support along the way. By order of travel, they include Hugh Alexander at the British National Archives, Sarah Rhodes and Dawn Young at the Bodleian Social Science Library, Oxford University, Christine Parker at the Faculty of Law, Melbourne University, Bob Nelson and the American Bar Foundation in Chicago, Doron Weissman in New York and Shaul Perry in Amherst, Claris Harbon in Montreal, Susan Silbey at MIT, Cesar Rodriguez Garavito and the Global Justice and Human Rights Program at the University of Los Andes in Bogotá, and Sol Picciotto and the kind people at the Onati International Institute for the Sociology of Law in Basque Country.

I met many heartwarming people along the road, all of whom facilitated my ability to think and write. I am grateful to Tania and Eitan Amir in London; to Laura Rico Pineres, Edna Martinez, and Diana Bocarejo in Bogotá; Dr. Swati Shirwadkar of the Department of Sociology at the University of Pune, India; and Gil Merom in Sydney. I owe special thanks to Carina Frahm and Bruno Hohmann, my wonderful neighbors, who sustained me throughout a four-day electricity blackout in my Bogotá apartment. It was into their electric outlets that I plugged my phone and computer, always accompanied by fine wine and meals, candles and torches, and a big welcome.

Some good friends and colleagues read drafts along the way. First and foremost was Christine Parker, who patiently read through early and advanced drafts and provided excellent comments. I also benefited greatly from the reading of draft chapters by Jean Comaroff, Tamar Elor, Daniel Monterescu, Kineret Sadeh, Nadera Shalhoub-Kevorkian and Susan Silbey. I owe many thanks to Mr. Moshe Feintuch and Mrs. Haya Ben Yaacov, at the Israel Electric Corporation's Haifa Archive, , who were always ready with an open mind to my queries and diggings. I am also most grateful to Nelly Varzarevski, at the Tel Aviv Municipality Historical Archive, for being an unending source of help, ideas, good will, and genuine enthusiasm for this project. I am grateful to colleagues who came forward with archival documents and valuable suggestions: Michael Birnhack, Leo Coleman, Yona Eliad, and Fredrik Meiton.

Three research assistants accompanied me along the way, each of them a treasure and a pleasure to work with: Tena Erez, who was there from the start, enthusiastic and inspiring; Mor Shilon, without whose help I would not have been able to find my hands and feet; and Chen Haklai, who walked me through the final stages of manuscript submission.

The comments of anonymous readers and the extremely valuable suggestions of my editor, Kate Wahl, ensured my ability to upgrade this manuscript and to be less anxious about the final product; for this I am very thankful. This research was generously supported by a three-year grant from the Israel Science Foundation (Grant No. 406/09).

Current Flow

Introduction

AMONG THE VARIOUS PHYSICAL INFRASTRUCTURES around us, the electric grid is critically foundational. From the production of electricity by diesel fuel, coal, nuclear, or hydroelectric power stations, to the running of electric current through high-, medium-, and low-tension wires, all the way to the electric appliances of the individual consumer—electricity is the quintessential infrastructure, feeding and sustaining other essential grids like water supply and irrigation, railway systems, and the Internet. Wherever we go and whenever we look, we see electric wires and their visible energetic products. They crawl under our feet, spring out of the ground, and cut across the sky. Although the electric grid is an old-fashioned infrastructure, consisting of concrete physical connections between one point and another, we cannot do without it—all other wireless grids rely on it. Imagine not being able to charge—for example, not being able to periodically insert your laptop or cell phone into a source of wired electricity. "It's outdated," complains one writer. "It isn't suited to meet our power demands." Still, we are absolutely dependent on it.[1]

Although we tend to think of the grid—a cobweb of wires a century and a half old—as a long-completed project, electrification is still an ongoing process. Grids never sleep; once they more or less cover any definite area, their tendency is to transcend the boundary they have just established and expand beyond it through outreach to yet unconnected areas or attachment to a neighboring grid. On the one hand, vast areas of the world are still unconnected to electricity—for instance, Sub-Saharan Africa, where most of the rural population is unwired. On the other hand, soaring demand prompts separate grids to join forces. Consider the recent trilateral agreement between Greece, Cyprus, and Israel to lay a 200-megawatt underwater cable that "will ultimately link Israel's electricity supply with that of the European Union."[2] From the outset, processes of electrification substantially "transformed the landscapes of the city, factory, home, and farm" (Nye 1990: 381). I should add that processes of electrification have also had the capacity to transform the landscapes of regions and countries and transnational space; because of their tendency to expand, they may even reconfigure the dimensions, coordinates and topologies of space itself.

This book is about electrification: the construction of a powerhouse and the assembly of an electric grid. It considers the materials the grid is made of—wires, poles, generators, transformers, current meters, and numerous other big and small devices—all participating in the generation, circulation, transformation, and distribution of electric current to and through multiple contact points. Many other connection types are involved in the process of bringing about wired electric light: "administrative," "legal," "political," "diplomatic," "imperial," "personal," "cultural," "financial," "ideological," and "technical." Yet such categorical designations are only analytic shorthand. None of these types and sets of connections are independent of any other. Each connection type invokes a former one or, at times, a new one, sometimes of an entirely different order. Ultimately, all of these connections and attachments come together to assemble a grid made of copper wires and poles; electricians, technicians, and engineers; consumers and officials; textual and graphic representations; and technical and legal documents.

I write this account as a sociologist, not as an electrical engineer or a historian of technology. My aspiration is to contribute to social theory, and my intention is to do so by probing into the electrification of 1920s colonial Palestine. My empirical case certainly aims to converse with other studies of Palestine and hopefully to enrich the understanding of its history. It also aspires to go beyond its idiosyncrasies to demonstrate how the trajectory of an electric network may yield relevant insights for sociohistorical research. Here is my argument: electrical connections participate in processes of group formation, take an active part in the performativity of social asymmetries; shape areas and regions and other spatial formations; and actively assemble, sustain, and enable taken-for-granted categories and dichotomies such as the private and public spheres.

Major works on the origins and diffusion of electricity have so far focused on Europe and the United States. This is not surprising. It was in the industrialized countries of Europe and in the United States that electricity "began": discovered, developed, and eventually widely transmitted and distributed. Thomas Hughes's *Networks of Power: Electrification in Western Society 1880–1930* (1983) set the tone, announcing a departure from technological determinism—that is, the assumption that technological innovations have a history of their own, independent of other social influences, and that they in turn function as a governing principle in shaping cultural and political processes. Hughes moved away from the tendency to account for technological successes and failures merely in terms of the cost, efficiency, and safety concerns that they satisfy (or fail to satisfy) on the road to fulfilling some preordained social needs.

Hughes launched a study of electricity that placed the "residuals" of political and economic factors at the center of inquiry. However, he did not simply move from technological determinism to "social" explanations for technological developments. Rather, he insisted on analyzing electrification in terms of a "seamless web" of connections and contributing factors (Hughes 1986; Nye 1990; Latour 2005: 81). Comparing the history of electrification in Germany, Britain, and the United States, he showed that politics and science; administrative, legal, and engineering concerns; businesspeople and entrepreneurs; inventors and investors; technical problems and engineering solutions; and regional geography and economic cycles were all involved in establishing the pace and form of electrification and its uneven development and rationalization in these countries.

To make sense of this seamless web, Hughes adopted an evolutionary approach to the history of electric systems (1983: 7–9): "Although the electric power systems described herein were introduced in different places and reached their plateaus of development at different times, they were related to one another by the overall model of system evolution"

(1983: 14). He assigned agency and concrete properties to electric systems: they had a will and a sense of direction; "young" systems behaved differently from "old" ones (1983: 15); they had a "style" (Nye 1990: 79).

All in all, Hughes identified four evolutionary phases in the career of electrical systems: "invention and development," "technology transfer," "system growth," and "substantial momentum" ("mass, velocity, and direction") (1983: 14–15; see Chapter 4 in this book). Within this history of system evolution, he was adamant in showing that progression (in each case and in comparative perspective) was never linear, that it faced setbacks and problems, and, most important, that it could not be explained only in terms of efficiency, rationality, and cost-effectiveness. Like Hughes, Mark Granovetter and Patrick McGuire analyzed the American electricity industry in terms of its "identifiable social networks" and showed that "the way the electricity industry developed was only one of several possible outcomes, and not necessarily the most technically or economically efficient" (1998: 148; see also Platt 1991).

It was only after solving key technological issues and stabilizing standards and models of electricity production and distribution that the electric grid fully arrived in countries outside the industrial core. Like telegraphy before it, wired electricity was often carried on and through the circuits of colonial rule. The diffusion of electricity in Europe and the United States presupposed a material and human infrastructure already receptive to the new technology—in other words, a relatively high degree of industrialization, relatively developed transportation systems, and considerable numbers of urbanites who could afford to be wired and transformed into private and commercial consumers. As the case of Palestine will demonstrate, colonies and other countries on the periphery of the industrial world lacked such infrastructure. Consequently, electricity had to discover viable directions of flow and create from scratch its sources of demand. On the one hand, electrification in the colonies actually took off straight from the third and fourth stages that Hughes assigned to electrical systems. On the other hand, the electric grid had to search for direction in the absence of the relatively obvious sources of demand that existed in parts of Europe and the United States.

Perhaps it is no coincidence that the few studies that consider electrification in former colonies tend to view the process through the prism of identity politics. Historian Moses Chikowero, noting the lack of research on the history of electricity in Africa, has studied the electrification of Bulawayo in colonial Zimbabwe (Rhodesia) from 1894 to 1939. Directly linking the process to "power politics," he shows that electrification was racially based and served the white settler community while excluding Africans (Chikowero 2007). And Rao and Lourdusamy, looking at electrification in the Madras Presidency of colonial India from 1900 to 1947, show that linguistics-based regional politics—not merely techno-geographical considerations—led to the creation of two separate electrical grids (Rao and Lourdusamy 2010).

The studies of Rao and Lourdusamy and Chikowero swiftly move from the technical to the political aspects of electrification. Both presuppose already stable and identifiable ethnic or racial groups that used electricity to further their goals and ends. This book diverts from this line of reasoning and insists on avoiding a categorical distinction between the technical and the political. Specifically, it does not account for electrification in terms of already established power politics and already consolidated national movements (i.e., Jewish-Zionists and Arab-Palestinians) pitted against each other. Rather, the purpose here is to explore how electrification "makes politics" rather than merely transmits it—how electrification participates in the formation of distinct ethno-national groups rather than simply reflecting it. This distinction is subtle yet crucial, and it requires further elaboration on the logic of inquiry.

Let us pose two questions: Is electrification only a technical process of sending electric energy from one point to another? Or is it also one of the many means by which dominant groups (economic elites, ruling classes, ethnic groups, etc.) enhance their power and consolidate their superiority? In conceptual terms, are we to treat the wires that transport electricity to various contact points as intermediaries or as mediators? As first explained by Bruno Latour, an intermediary "transports meaning or force without transformation: defining its inputs is enough to define its outputs." Mediators, on the other hand, "transform, translate, distort, and modify the meaning or the elements they are supposed to carry"; "their input is never a good predictor of their output" (Latour 2005: 39).

Let us now reconsider the two questions. With this terminology in mind, it may be plausibly argued that *both* tacitly presuppose that electric wires function as intermediaries. A positive answer to the first question is almost trivial. Once the "war of currents" was decided, the whole point about the flow of (alternating) electric current became the fact that it remained stable throughout its long-distance circulation.[3] Knowing the input of electricity at the source (the generating facility) accurately predicts the output at the other ends of the grid.[4] A positive answer to the second question is less trivial, but if such an answer is chosen, it also treats electrical wires as sociologically neutral; electric wiring is only a way to consolidate a social force that is already there. In this sense, the electric grid does not matter much; just like other machines, procedures, rituals, or even parliamentary politics, electric wires are just a medium for transmitting—embedding in electric current—the force of some social group (a nation, a class, a criminal organization, etc.) from one location to another or from one position to another.

The path of the present study leads elsewhere. It suggests that electrification—specifically, the concrete material infrastructure that enables it—deploys numerous mediators. The political and economic circumstances behind it are not necessarily a good predictor of outcomes. The process of electrification can be neither reduced to its technical elements nor fully accounted for in terms of already existing political, cultural, or economic factors that determine its nature and scope. And just as electrification cannot be decomposed into its distinct political and technical aspects (i.e., science and society), so it cannot be treated as having only "social origins and effects" without being treated as "social" in and of itself.

The laying of a grid does something more than transmit electricity by connecting wires to a source of power. It does more than assemble a material infrastructure. In fact, it connects sketches, diagrams, maps, and contracts to lamps, transformers, poles, and current meters. It brings together investors, entrepreneurs, electricians, engineers, lawyers, government officials, statisticians, and expert advisers. It welds laws and regulations to commercial investments and political aspirations. It generates and affirms "identifiable social networks" (Granovetter and McGuire 1998: 148). It creates areas of coverage that translate into separate regions. In these ways, the grid is a maker of groups and a generator of political and economic difference among groups and individuals.

The study of an evolving grid should also be a good way to connect otherwise seemingly unrelated objects of study. Studying "technopolitics" in Egypt, Timothy Mitchell observes: "There are studies of military tactics, irrigation methods, Anglo-Egyptian relations, hydraulic engineering, parasites, the sugar industry, and peasants. But there are no accounts that take seriously how these elements interact" (2002: 27). Their heterogeneity, he writes, seems to resist explanation, and this resistance in turn "may have something to do with the mixing of natural and social worlds" (2002: 27), with each subject area seemingly locked into a separate science. I treat the electric grid and its dynamics of growth as a social assemblage also because it provides an opportunity to bring together some seemingly unrelated issues such as ethno-national relations, irrigation methods, municipal governance, colonial rule, orange groves, group formation, and railway networks.

The grid, in other words, cannot be understood only in terms of the context within which it is made possible. The method of inquiry that I propose here does not assign "context" a determining role in accounting for the general direction or order of events. The conceptualization of context as container within which things happen is replaced here with an imagery of a "flattened topography" (Latour 2005: 174). The unleveled landscape of concrete actors embedded in a general frame (i.e., big and small, high and low) is transformed into a terrain of equals. This method of inquiry does not ignore "context." Rather, it traces the actions and movements that link the sites of big and small so as to make them commensurable: if context matters, let it make itself present in and through these movements.

In studying grids, one direction follows the premise that "every new technology is a social construction and the terms of its adoption are culturally determined" (Nye 1990: 381). In methodological terms, this direction includes the experience and perspective of users in accounting for the spread of electricity and the expansion of grids. Nye shows that electricity was initially introduced as a phenomenon of the urban public sphere and was perceived as a spectacle of modernity in theaters, fairs, and newly electrified streets (on the distinction between "lighting of festivity" and "lighting of order," see Schivelbusch 1995: 137 and Chapter 3 in this book). Kline shows that the extension of electricity from the city to the countryside, presumably introducing urban modernization to rural areas, led to a complex cultural dialogue between urban and rural people. Rather than unambiguously embracing these hallmarks of progress, farmers resisted, modified, selectively used, and assimilated the new technology "into existing social patterns" (2000: 269).

Both Nye and Kline, then, deploy a user-centered approach to show that cultural factors influenced the reception of electrical power in ways that overrode utilitarian considerations.[5] They find that, at least in its early phase but sometimes extending well into the 1930s and 1940s, the direction and shape of the grid widened the gap between rich and poor and between urban and rural. So the grid "does" something, and the sociology offered in this study traces the way things become attached and connected and, in the process, produce categorical and epistemological differences and distinctions.

The method of inquiry adopted in this book concerns itself with tracing the expansion of the electric grid from the point of view of electric wires and poles: What were the directions in which electric current flowed? Who was wired? What was connected? Following the wires, various chapters of this book shed light on the ways in which electrification marked, affirmed, and produced a variety of social differences: between Arabs and Jews, between urban and rural, between private and public, between state and market, and between industry and agriculture as components of an "economy."

The book follows in the footsteps of, and uses some tools made available by, Latour's "sociology of associations" (2005: 7). The underlying theoretical principle at work here is that no such autonomous domain that can be properly labeled "social" exists distinct from other domains (political, economic, technological, etc.) and that no such thing as a "social context" exists as a hovering universe that may explain—more or less successfully—various presumably "nonsocial" affairs or entities (e.g., a technological development, a mathematical formula, an earthquake). The "social," as Latour articulates it, is "a trail of associations between heterogeneous elements" and it is "what is glued together by many other types of connectors" (2005: 5).

A Note on Method; Or, What Does ANT Have to Do with It?

Actor-network theory (ANT) began in studies of science and technology (Law 1992) and is still strongly associated with these domains. Yet as a sociological method of inquiry and, moreover, as a sociological sensibility and a guideline for sociological reasoning, it should yield useful insights for sociological inquiry. ANT-oriented or ANT-inspired ideas have spilled into studies of identity and subjectivity (Moser and Law 1999; Mol 2008); economic markets (Callon 1999); urban settings (Stalder 2006; Valderrama and Jorgensen 2008); media, communications, and information systems (Stalder 2002, 2005); passion and addiction (Gomart and Hennion 1999); and ethnography (Henare et al. 2007; Beaulieu 2010). ANT approaches have also made steady progress in sociological political theory, specifically political economy and theories of the state. Lawrence Busch and Arunas Juska (1997) rely on ANT to question traditional approaches of political economy to the globalization of agriculture. Others use it for revisiting sociological theories of the state, political economy, and democracy (Stalder 2006; Whitehead 2008; Callon et al. 2009; Passoth and Rowland 2010; Carroll 2012).

In this study, it is not state theory or political economy but ethno-national relations that are the center of attention. The purpose is to track down a process of electrification under the auspices of a colonial government and show how it produced and affirmed ethno-national distinctions. It was not the social or economic or political forces "behind" the process of electrification that best accounted for its consequences. Rather, the flows and connections of the newly created and rapidly expanding grid underwrote processes of ethno-national distinctions and were active in assembling these very social and political forces that seem to have preceded it. This logic of inquiry renders ANT particularly attractive. However, there are a number of misconceptions about ANT that need to be spelled out and addressed by way of explaining the present study.

The first notion (or hope) to be discarded is that ANT is a theory about society or agency or subjectivity or relations or network formation—or about anything at all. ANT is not a theory: "Theories usually try to explain why something happens, but actor-network theory is descriptive rather than foundational in explanatory terms, which means that it is a disappointment for those seeking strong accounts. Instead it tells stories about 'how' relations assemble or don't" (Law 2008: 142). ANT does not compete over any single terrain with other theories. Rather, it is "a diaspora that overlaps with other intellectual traditions" (Law 2008: 142). It is a toolbox, perhaps best captured as a "sensibility." Thus, ANT does not and cannot offer a theory about electrification or the formation of electric networks, systems, and industries. It also does not and cannot outline a theory about the formation of separatism in Arab-Jewish relations in Palestine. This study follows the idea that ANT is a method: "It says nothing about the shape of what is being described with it" (Latour 2005: 142). Even as a method, let alone a sensibility, "purity tests would seem antithetical to ANT. . . . Those most committed to ANT seem least invested in policing its use" (Carroll 2012: 9). I fully embrace these statements as well.

A second issue that must be briefly clarified has to do with the status of nonhumans in ANT. Having its roots in science and technology, ANT acquired a reputation as a theory about the role of objects and nonhumans in social affairs. This reputation is not unfounded: "ANT studies began in scientific laboratories, documenting *knowledge in the making*—the often ad hoc constellation of people (e.g., scientists, patients, grant committees, etc.), things (e.g., beakers, measurement tools, grant proposals, laboratory benches), ideas (e.g., Newton's Laws), resources, etc., through which object X or theory Y are produced" (Levi and Valverde 2008: 807). There is absolutely nothing wrong in seeing fascination with ANT in the way objects participate in the assembly of associations, act in the constitution of networks, or bring about change in some stable state of affairs. Yet this way of thinking about it risks celebrating the exotic while missing the heart of the matter. ANT is a notoriously slow and laborious way of describing how matters or changes or issues or states of mind or worldviews or anything else come together—and, more often than not, only tentatively, uncertainly, and in flux. In the process, indeed, the method of inquiry is sensitive to the role of nonhumans as connectors and translators, as elements of a network, as "immutable mobiles" whose stable and unmodified circulation enables networks, or as boundary objects "that can change depending upon how they relate to other objects and actors" (Carroll 2012: 8).

The principled methodological decision in this study has therefore been to follow the wires, to let the poles act, and to trace the flow of the newly distributed electric current in Palestine to private residences, water wells, workshops, orange groves, streetlights, and designated townships. At my disposal were materials from six archives in Israel and the United Kingdom. Three of these have proven particularly rich and useful: the records of the Colonial Office in the British National Archives in London, the records of the Jaffa Electric Company (later absorbed into a newly created Palestine Electric Company) in the archive of the present-day Israel Electric Corporation, and the records held at the Historical Archive of Tel Aviv Municipality. They correspond to the most important triangle of relations underlying electrification in Palestine: the British government in London and the British authorities in Palestine, the Jaffa Electric Company that was licensed by the British to undertake the project, and the Jewish town officials who pressed for the electrification of Tel Aviv (the first town to be wired in Palestine in 1923, followed by Jaffa later in the same year).

The archival records allowed me to trace intensive exchanges of letters and telegraphs; contracts and drafts of contracts; expert opinions; maps and sketches; technical specifications; and policy memoranda that circulated between London, Jerusalem, and Tel Aviv. Gathering them allowed me to re-enact the construction of the first electric grid in Palestine, covering the District of Jaffa and soon spreading to other areas of the country. The documents tell a story of assembly: electrical machinery linking up with investors, engineers, policy makers, laborers, experts, officials, national ideologies and organizations, colonial interests, and ordinary people to create a power station that produced electricity and distributed it through a grid of wires. Hughes (1983) refers to such projects as "networks of power," and this label merits a comment about a third possible misconception concerning ANT.

The electric grid is a technical network: it is made of physical elements that constitute a set of connected nodes, and it circulates electric current in a relatively patterned and stable way. It is also possible to describe it as a social network, in the straightforward sense that beyond its physical elements it includes an interdependent set of human (or institutional) actors interacting with varying degrees of intensity. However, such characteristics do not make electric networks particularly "suitable" for ANT study and analysis. There is nothing about ANT that should privilege it in the study of networks of whatever type. The "network" in the ANT acronym stands for something different that, at the very least lies in the fact that the actor network must be constantly operating. It cannot be permanently fixed because it is only a network when in action—minute, trivial, and often unpredictable actions enable it, and it ceases to be once this essentially messy and heterogeneous work stops.

Helene Buzelin writes that, like all networks, the actor network is "defined in terms of connections rather than distance and in terms of extension rather than scale. Yet actor-networks should not be confused with technical or social networks." Unlike social networks, actor networks encompass "anything that can induce" an action (including nonhumans). And while technical networks (electronic, rail, etc.) appear as a given structure that can be extended—hence as something that can be mapped—"actor networks can only reveal themselves when activated" (2005: 197).

Thus, matters such as "action" and "performativity" are crucial for distinguishing actor networks from other networks. Still, the very juxtaposition of actor networks and other networks and the singling out of their differences may create the false impression that actor networks are just another type of network, which brings us back to what is meant by emphasizing action. Latour explains this point to a fictional doctoral student: "You should not confuse the network that is drawn by the description and the network that is used to make the description"; more instructively: "With Actor-Network you may describe something that doesn't at all look like a network—an individual state of mind, a piece of machinery, a fictional character; conversely, you may describe a network—subways, sewages, telephones—which is not all drawn in an 'Actor-Network' way" (2005: 142). Technical or social or socio-technical networks are tangible objects of description (however "socially constructed"), while ANT is just a method for drawing a description; it specializes in tracing connections, associations, and modes of movement among actors. This fundamental distinction is particularly important for clarifying the present study, when some of the sensibilities associated with the tool (ANT) are used for describing a network (an electric grid).

Equipped with such ideas about the character and merit of ANT, the purpose here is to study the electric grid not because "it is an actor network" but because an actor-network approach may be useful in making sense of the grid as a social assembly in and of itself. This cannot be understated: the grid is "social" not only because it is made of "social stuff" and has "social effects" but also because it performs and enables (or disables) social formations through the physical connections it establishes and its attachments to other entities or matters of concern. The analysis of an electric grid in such terms—paying particular attention to the initial phase when it is constructed and begins to expand—should be useful both for treating the case of Palestine and more generally for engaging historical and political sociology with hardware and infrastructures.[6]

Palestine: Some History and Historiography

British control of Palestine began toward the end of the First World War and ended in May 1948. The British Army first entered Palestine from the south, climbing its way up from Egypt and pushing the armies of the crumbling Ottoman Empire further north. British forces captured Jerusalem in December 1917 and completed the conquest of Palestine and Trans-Jordan in 1918.

In November 1917, Foreign Secretary Arthur James Balfour sent a letter to Baron Rothschild, a leading Jewish-British Zionist, and through him to the Zionist Organization of Britain. Soon to become widely known as the Balfour Declaration, this letter promised that "His Majesty's Government view with favour the establishment in Palestine of a national home for the Jewish people, and will use their best endeavours to facilitate the achievement of this object, it being clearly understood that nothing shall be done which may prejudice the civil and religious rights of existing non-Jewish communities in Palestine" (quoted in Gilbert 2007: 27).[7] The Balfour Declaration was incorporated into the preamble of the "Mandate of Palestine" granted to Great Britain by the League of

Nations in 1922, transforming the promise to facilitate a Jewish National Home as an internationally sanctioned obligation.

In 1920, the British military regime was replaced by a civil administration that established its rule in accordance with the blueprint of a crown colony (Biger 1983), relying on experienced civil servants and colonial officers who moved to Palestine from colonies such as Cyprus, Mauritius, Rhodesia, Malta, and Sudan. This administration created government departments (education, health, finance, etc.), imposed and regulated customs and taxes, established and managed the Palestine Railways and the Palestine Post, reformed and developed a courts system and a police force, reorganized Palestine as a coherent geographical entity, built communications and transportation infrastructures, and in general assumed and executed broad state functions (Samuel 1957; Shapira 1989; Reuveny 1993; Segev 1999; El-Eini 2006).

One of the earliest decisions of the newly instituted British administration was to grant Pinhas Rutenberg—a Jewish entrepreneur enjoying the support of Zionist institutions and financiers—an exclusive license to electrify Palestine. Rutenberg was thus what the British authorities in Jerusalem and London considered a "concessionaire."[8] The granting of concessions, in and of itself, was not novel or unusual but was an essential and widespread instrument of colonial rule.[9] Based on liberal economic theories, concessions were used as a system of indirect rule, limiting and shaping the operations of markets in the colonies while often striving to achieve or maintain political ends.[10] They were also regarded as a tool for advancing imperial trade, boosting national industries, and guaranteeing British labor at home. On the one hand, concessions relieved the colonial state and British taxpayers from having to finance public works and development projects in the colonies. On the other hand, British officials and agents of the Crown justified them as efforts to modernize and develop the colonies for the benefit of the colonized (Sunderland 2004: 237–239).

Individual entrepreneurs and organized syndicates thus scrambled for concessions all over the British Empire: in Africa, Asia, the Middle East, Latin America, and the Caribbean (Moberg 1996; Barton and Bennett 2010). However, Palestine was unique in that British companies and investors did not consider its electrification as potentially lucrative (Smith

1993: 132). Palestine in the 1920s lacked all the essentials that were deemed necessary for profitable electrification: large urban centers, a mass of people who could afford to pay for the new technology, developed urban and interurban transportation systems, and power-hungry industries. However, as was sometimes the case in other colonies—where intricate and mutually beneficial relationships existed between colonial governors, concessionaires, and British colonial officers (Sunderland 2004)—electricity concessions were strongly advocated by the High Commissioner for Palestine and sanctioned by the Colonial Office in London under a general theory of economic development that would pacify Arab-Jewish relations.

Pinhas Rutenberg, in spite of strong opposition from the Arab leadership of Palestine and some British politicians, was thus granted exclusive rights to generate, distribute, and sell electricity in the District of Jaffa and later in the whole of Palestine and Trans-Jordan. These rights were delivered through two detailed contracts—the "Auja Concession" (the Yarkon river near Jaffa) and the "Jordan Concession" (the intersection of the Jordan and Yarmuk rivers south of the Lake of Galilee)—that were worked out by a team of government officials, technical advisers, and lawyers on behalf of both parties.[11] Both concessions expected electricity to be produced by hydroelectric technology. The Auja Concession for the Jaffa District was formally signed on September 12, 1921; the Jordan Concession for the whole of Palestine was signed much later, on March 5, 1926.

Apart from strong Arab opposition to the granting of concessions to Rutenberg, the British had to deal with a conflicting demand that was rooted in Ottoman rule. Agreements for the supply of water, electricity, and the construction of a tramway system—all under concessions granted by the Ottoman authorities—had been signed by the city of Jerusalem and a Greek citizen by the name of Euripides Mavromatis on January 27, 1914. Work under these concessions had not begun and, by the end of the war the British occupying forces refused to recognize their validity. Mavromatis claimed that his concessions overlapped with the Auja Concession that the government had awarded to Rutenberg in 1921 and that he had been deprived of his legal rights. Herbert Samuel, the High Commissioner of Palestine, in a letter to Churchill, confidently dismissed these claims. He wrote that the concession for Jerusalem was granted to Mavromatis by the Ottoman authorities in his "character of an Ottoman subject. It would, therefore, appear to be annulled in virtue of article 315 of the treaty of peace with Turkey. Mr. Mavromatis indeed now claims to be a Greek subject, but even if he could substantiate this claim the concession may be declared null and void in virtue of article 314 of the treaty. It is, therefore, considered that the grant of the concession to Mr. Rutenberg is not affected by this claim."[12]

The opinion of the High Commissioner turned out to be a grave error, one that would implicate the British in costly and lengthy legal battles. Mavromatis enlisted the Greek government on his side and took the case to the Permanent Court of International Justice in The Hague. In a preliminary ruling in 1924, the Court established its jurisdiction to decide on the merits of the dispute. Subsequent deliberations in the Colonial Office concluded that "the result of the decision now is that we shall have to fight the case on its merits as far as the Jerusalem concessions for Electric Power and Tramways and Water Supply are concerned."[13] In a follow-up ruling in 1925, the Court decided that, according to the principle established by the 1923 Protocol of Lausanne, the concessions granted by the Turkish authorities before October 29, 1914, were to be maintained.[14] Specifically, the Court upheld the British position that Mavromatis had no rights with respect to electric works in Jaffa, but upheld the Greek position with respect to his concession rights in Jerusalem.

The result was that Jerusalem and its immediate vicinity became an electric enclave. Elsewhere in Palestine, the Jaffa Electric Company was established to fulfill the obligations of the Auja Concession. It was later absorbed into the newly founded Palestine Electric Company, which was created to meet the obligations of the Jordan Concession.[15] As I show elsewhere in this book, concessions—translated into actual grids—can shape boundaries and regions, and sometimes they have the capacity to set populations apart well ahead of other political plans of partition and segregation.

In 1922, roughly 83,000 Jews lived in Palestine, compared to 660,000 Arab Muslims and Christians (a 1:8 ratio). By 1936, 400,000 Jews lived in Palestine alongside 740,000 Arabs, dramatically changing the demographics of the country (roughly a 1:1.8 ratio). Yet the demographic transformation tells only a small part of the story. Over the course of these fifteen years, the Jewish community in Palestine gained confidence in its ability to sustain itself independent of the Arab majority and without the protective umbrella of the British. The episteme of separatism was born—the conviction that the two societies could not and should not share one national umbrella and that a spatial segregation between Jews and Arabs was not only desirable but also viable and sustainable.[16]

In 1937, in the midst of a violent Arab-Palestinian revolt against British colonialism and its reputed support of Zionism, a Royal Commission of Inquiry (the Peel Commission) suggested for the first time a "partition plan" for Palestine, expressing despair at the possibility of peaceful coexistence and recommending the territorial division of Palestine between Arabs and Jews (Kimmerling 2004; Dothan 1979).[17] It took another eleven years until the episteme of separatism reached its violent conclusion in the war of 1948: the ethnic cleansing of Palestinians from those parts of Palestine that were to become Israel and the creation of the still persisting problem of Palestinian refugees (Morris 1998, 2004).[18] It is the very same episteme that is alive and well today.

So the "end" is well known, and it is through these charted waters of studying and theorizing the roots of separatism (Smith 1993) that the present inquiry sails. Studies of Palestine, at times approving and at times critical, at times in celebratory tones and at times in tones of mourning, seem to agree that in the course of three decades of British rule the Jewish population of Palestine managed to create "a state within a state," one that gradually disengaged from the Arab population and asserted its increasing autonomy from general prevailing sociopolitical conditions. Accounting for this process, one line of reasoning and inquiry emphasizes the crucial role of the British in enabling the success of Zionist "nation building"/colonizing. Some observers broadly suggest that the creation of the Jewish State in 1948 could hardly be imagined if not for the preceding thirty years of British rule over Palestine. Israel came into being only after the Jewish population gathered sufficient strength to overwhelm the Arab-Palestinian community, and this acquisition of power would have been impossible without British support (Huneidi 2001; Segev 2000; Khalidi 2006).

Historian Anita Shapira writes that most of the young Jewish immigrants who came to Palestine in the early 1920s would have been compelled to leave it "if it had not been for Herbert Samuel (the first British High Commissioner of Palestine), who initiated basic infrastructural development projects" that provided them with gainful employment (1992: 131). Amos Nadan (2006) has analyzed the Arab-Palestinian peasant economy under British rule, showing that Jewish agricultural developments did not benefit the Palestinian-Arabs and that colonial efforts to reform cultivation and the use of credit were premised on British assumptions of Arab backwardness and irrationality.

Barbara Smith locates the roots of separatism in British economic policies that benefited the Jewish population. She identifies the preferential treatment of the Jewish population in immigration and land policies, in the protection of Jewish industry, and in fiscal policies and public finance. She also singles out the advantageous consequences of granting Zionists various types of monopoly concessions. Paying particular attention to electricity concessions, Smith decisively concludes that electrification, under British auspices, was "fundamentally a Zionist project" (1993: 119).

Some of the findings in the following chapters suggest otherwise. While it is certain that this project could not have taken place without the political support of the British authorities, and while it is certain that it was financed by Zionist institutions, the present study doubts the premise that the electric wires were just intermediaries for transporting the colonial might of British-protected Zionism.

Another common theme in tracing the roots of separatism is a focus on the self-help activities of the Jewish population: protagonists who took their fate into their own hands and managed to create "a state within the state" while overcoming both Arab resistance and hostile British colonizers.[19] This focus mainly emphasizes the self-organization capacities of Zionism—namely, the economic, political, cultural, and military institution-building drive of the Jewish population during the Mandate years. Economic historians and political economists have thus studied and documented the emergence of a "divided economy," a "dual economy," and a "split labor market" in Mandatory Palestine (Metzer and

Kaplan 1985; Shalev 1989; Kamen 1991; Metzer 1998). Quite a lot has been written about the decisive role of organized Jewish labor and its political and welfare institutions, as well as the role of public capital, in mobilizing Jewish nation building/colonizing. This emphasis is not unfounded. Zeev Sternhell, writing critically about the ideology of the Zionist labor movement, convincingly shows that it invested heavily in the creation of highly disciplined statelike institutions. Politically organized labor, he suggests, was about state building much more than about social reformation (Sternhell 1995).[20]

Less has been written about the role of Jewish private capital in enabling the "state within a state" and its subsequent politics of separatism (Shamir 2000, 2001).[21] The flow of Jewish private capital into Palestine was always greater than the flow of public capital (Kimmerling 1983). The two biggest waves of Jewish immigration to Palestine in the mid-1920s and mid-1930s consisted of commercial classes with little or no Zionist zeal (Giladi 1973; Ben Porat 1999), and it is practically impossible to imagine nation building/colonizing without the Jewish urban settling in and around Tel Aviv. Not only Jewish industry (Beilin 1987) but also the cultivation of oranges was a business of "planter-capitalists" especially along the coast and around Jaffa and Tel Aviv (Karlinsky 2000; see also Chapter 5 here). In fact, early expressions of "separatism" may already be located in the 1920s precisely with respect to the activities of private Jewish industry and agriculture.

The social sciences, write Law and Urry (2004), do not merely describe the world; they also make social realities and enact social worlds. In Palestine, "social-scientific separatism" emerged in the 1920s in the work of economists and statisticians who assembled data that validated the existence of a distinct "Jewish economy" and enacted the reality of two separate ethno-national communities. Assuming an all-Palestine perspective of British policies and the country's economy, and making a special effort to collect data about both Arab and Jewish economic activities, latter-day analyses also affirm the emergent reality of a divided economy (Sussman 1973; Metzer 1982, 1998; Gross and Metzer 1993; El-Eini 1996, 1997, 2006). Some sociologists translate this story line into a full-blown "dual-society" paradigm:

> Two populations so different from one another can exist in the same territorial framework in only one of two ways. The two populations can be aligned in the same hierarchy, with the more "developed" population (according to generally accepted indices of modernization) located in the upper stratum and the less developed in the lower stratum. Or, each population can maintain a hierarchy of its own, which implies that the populations are, to a considerable extent, segregated from each other. The first pattern often developed in areas of European migration and settlement and in colonial societies, such as South Africa and Rhodesia. In Mandatory Palestine, however, a social structure of the second type developed, a circumstance without parallel in the annals of European settlement in the underdeveloped colonial territories. (Horowitz and Lissak 1978: 6)[22]

Other studies of Palestine question the seemingly neat character of the dual-society paradigm, documenting not only separation and segregation but also exploitation and subjugation. A relational approach to the study of Palestine has emerged in response to this paradigm, either documenting instances of Jewish-Arab cooperation or simply emphasizing the mutually constitutive dynamics of interaction that provides groups with "others" against which they assert a distinct identity. Either way, the relational approach critiques the paradigm for portraying "the Arab and Jewish communities in Palestine as essentially separate, coherent, and self-evidently cohesive entities that developed along entirely distinct paths" (Lockman 1995: 212; also see Roberts 2011; Bernstein 2000; Kimmerling and Migdal 2003).

Earlier in this introduction I explained that the methodological principles that guide the present inquiry assume that group formation is a process "in action": rather than assuming coherent entities that are already there as determining forces, the study of electrification seeks to account for the ways in which this seemingly technical process actively shaped and contributed to the assembly and performance of such ethnonational formations and, arguably, contributed to a deepening divide between them. This methodology, in short, runs on a collision course toward studies that assume "self-evidently cohesive entities that developed along entirely distinct paths" (Lockman 1995: 212).

Of course, it would be disingenuous to claim that Arabs and Jews did not exist as different groups prior to electrification or that the decade of the 1920s was not already seeing its share of national tensions between them. Rather, the point is that collective identities and differences between groups have to be constantly maintained, affirmed, performed, made visible, and executed. At stake, then, is whether the electrification of Palestine enabled and set into motion a deepening ethno-national divide between Palestinian Arabs and Jews under British colonial rule. And this research effort is particularly challenging if it does not a priori equip itself with the idea that electrification was determined by the fact that the concession was given by the British to a Jewish entrepreneur who enjoyed the financial backing of Zionist institutions.

"The most important proposition revealed by history," writes novelist Haruki Murakami, is that "at the time, no one knew what was coming" (2011: 11). This is an important reminder that bears on my decision to focus only on the first years of electrification in the 1920s, before the completion of the larger hydroelectric power station on the Jordan River in 1932. I cannot emphasize enough the importance of this early period. By the mid-1930s, and definitely in the 1940s, the idea of an unavoidable need to separate Arabs and Jews in Palestine had become hegemonic in the minds of many and was already firmly embedded in British policies. The whole point about the 1920s is that the material consolidation of the grid (with its nascent differences and distinctions) preceded and "assembled" part of what was to come.

The electrification of Palestine during the 1920s took place when the possibility of territorial and political separation between Arabs and Jews was for all practical purposes inconceivable. Many roads were opened in 1920s Palestine, and the bitter consolidation of Arabs and Jews as two ethno-national groups pitted against each other was still far from becoming an inevitable, irreversible trajectory. Many Arabs had deep misgivings about the Zionist endeavor in Palestine in the aftermath of the Balfour Declaration, but there is a critical difference between everyday-life practices of ethno-national distinction and an episteme of separatism—that is, the conviction that two ethno-national groups are fundamentally incompatible and must be politically and geographically separated.

At the time, Arab-Palestinians warned the British that the electricity concession would give Zionism unfair economic advantages and that the whole process of granting concessions exposed British favoritism toward Jews, giving the latter "a stranglehold on the economic life of Palestine and Transjordania."[23] These concerns were valid. A monopoly on an essential infrastructure such as electricity has the potential to yield considerable advantages to those who are in control of the switchboard and to those consumers (residential, municipal, industrial, and agricultural) who are able to join the grid. Simply put, those who are connected to an electric grid enjoy possibilities that are denied to the unconnected or to those yet to be connected.

The electric grid in Palestine, as elsewhere, began its career in the city, lighting the streets, providing water, attaching itself to residential homes, and, wherever possible, linking up with industry. From the point of view of an electric grid seeking to expand, Arab municipalities and individual Arabs in urban centers like Jaffa and Haifa, and in smaller towns like Acre, Ramleh, and Nazareth, were important as prospective consumers as much as were Jews in Tel Aviv and other locations. And from the same grid-centered point of view, at least for a while electric current flowed in the direction of state-run enterprises and installations (e.g., railways) and thereby sought to secure the government as a heavy consumer. The electrification of Palestine, sanctioned by the Mandatory State, was a countrywide commercial enterprise and, according to the British, was moreover intended to further the economic development of the country precisely in order to pacify relations between Arabs and Jews. In principle, then, electrification in Palestine seems to be a strong case of a process that sought to appeal to the general population and transcend whatever boundaries existed between Arabs and Jews by the sheer force of a grid whose raison d'être was to expand. It is a case that offers an opportunity to assume an all-Palestine countrywide view and to explore how it still may have ended up accelerating the ethno-national divide it was designed to neutralize.

Instead of privileging the primacy of Jewish-Arab relations in understanding electrification, I examine how differences between Jews and Arabs were performed and enacted through the material process of laying out cables and electric poles. Group formation and ethno-national difference become here the dependent variables to be accounted for rather than the explanation for the process. And perhaps the most important point in concluding this introduction is this: trying to deploy a sociology of associations, this study aims to show that the "how" of electrification was by far more telling than the assumption that it was the means for translating Anglo-Jewish political power into economic power and, moreover, a deliberate move on the teleological road to separatism. It is in the details of "how," with all of their unintended consequences, that the following study is embedded.

A Word on Structure

Chapter 1 sets the stage for the book by looking at the construction of the grid's source—namely, the first electric power station on a piece of land between Jaffa and Tel Aviv. The first part of the book includes two chapters that should go under the heading of electric supply for "light," as they focus on the flow of electric current to streetlights and to private consumers who mainly used electric current for lighting their homes and businesses. The second part of the book should be read under the heading of electric supply for "power," as it includes two chapters on the attachment (or nonattachment) of the grid to railways, industry, and agriculture. The book properly ends with a brief conclusion.

I also offer another structure, in which Chapter 3 stands at the center, right between the first and second pairs of empirical chapters. This is the only chapter in the book in which Jewish-Arab relations do not occupy any significant part of the analysis. What may at first look like a diversion is in fact important to the book's overall posture. It is precisely because this chapter is devoid of the ethno-national theme that it helps to clarify what I mean when I speak throughout about the sociality of the grid: its extension works through a distinction between the "general distribution system" and "individual connections." The very work of extension simultaneously performs the divide between "public" and "private." Within the "private" sphere, the measurements of electric consumption enable the compilation of data according to categories of consumers and levels (or classes) of consumers. Both cases illustrate the way social difference, and principles for making social distinctions, reside in and on the grid.

1

The Jaffa Powerhouse

ON JUNE 10, 1923, a ceremonial photograph (Figure 1) was taken at the newly built powerhouse of the Jaffa Electric Company (hereafter the Electric Company or the company). Built in the best and latest high-tech architectural style, the powerhouse was located on a 5-acre (20-dunam[1]) lot somewhere along the eastern part of the loosely defined boundary between Jaffa and the small Township of Tel Aviv. Electricity first flowed through the electric grid to the two main streets of Tel Aviv on this day. Carried through low-tension cables, electrical power traversed from the transformer to 40-, 100-, and even 200-"candle-power" (watt) streetlamps. In the center left of the photo stands Pinhas Rutenberg, entrepreneur and chairman of the board of directors of the Jaffa Electric Company, dressed in spectacles and an open black jacket and tie.

The ceremonial lighting of the two main streets of Tel Aviv realized the first operational results of the Auja Concession, which had authorized Rutenberg to generate electricity by means of hydroelectric turbines that would exploit the water power of the small Auja River (the Yarkon in Hebrew) to supply electricity to Jaffa (1923 population: 55,000), the smaller neighboring town of Tel Aviv (1923 population: 16,000), and other locations within the bounds of the administrative District of Jaffa. Article 3 of the Auja Concession stipulated that the concessionaire would build a

FIGURE 1 *Pinhas Rutenberg and staff at the opening ceremony of the Jaffa Electric Company powerhouse, June 10, 1923*

SOURCE: BNA CO 1069/731.

NOTE: This photograph is part of an album documenting the construction of the Jaffa powerhouse. The photographs were all taken by Z. Brill, a professional photographer who was hired by the electric company for that purpose. The album was presented to the governor of Jaffa and currently can be found intact in the British National Archives.

hydroelectric "generating station or power house in the neighborhood of Jerishe to be called the Yarkon Power House."[2] Among many other things, the concession created an obligation on the part of concessionaire to complete all works and to begin operations by September 1923.

Yet the plan to generate electricity by hydroelectric means never materialized, and instead Rutenberg steered the company to design and build a powerhouse that produced electricity by means of diesel-fueled engines. The powerhouse was built miles away from the original designated location, in an area between Jaffa and Tel Aviv (Ram 1996). Rutenberg justified the shift in location and technology in terms of the political conditions pertaining to the electrification of Palestine. He explained that he was diverted from his original plan by the unreasonable financial demands of recalcitrant Arab landowners in the Auja basin, who had joined forces with an all-Arab collective struggle against the British decision to grant Rutenberg his concessions. The British authorities accepted this explanation, as have latter-day historians.[3]

Though British electric companies did not foresee any commercial value in the electrification of Palestine, the decision to grant the concessions to Rutenberg was nonetheless a product of imperial policies. The British had taken Palestine from the crumbling Ottoman Empire toward the end of World War I, and their foothold in the region received international backing through colonial agreements with France and later through a Mandate of the League of Nations. The Mandate expressed the British commitment to promote the necessary conditions for establishing a "Jewish National Home" in Palestine, along with a general obligation to develop the country for the benefit of all its inhabitants. The commitment to promote a Jewish National Home was particularly significant considering the demographic profile of the country at the time: 643,000 Arabs and 84,000 Jews according to the 1922 census. The grant of the concessions to Rutenberg (and to the Zionist institutions that provided financial backing to the electric companies he established) functioned as an economic medium for advancing imperial interests: they were portrayed by the British both as means for economic development of the country (and hence an instrument for pacifying ethno-national tensions) and as a market-based means for promoting the Jewish National Home without resorting to controversial political steps.

British hopes to neutralize the political implications of granting the concessions to Rutenberg were disappointed. The political and financial backing for the electrification plan came almost exclusively from Zionist organizations and Jewish philanthropists—the London-based Jewish Colonial Trust and the Palestinian Jewish Colonization Association (funded by Rothschild). Rufus Isaacs (Lord Reading) sat on the board of the Palestine Electric Company as the representative of several British trust associations, and other notable British Jewish funders included Sir Alfred Mond (Lord Melchet) and Michael Nassatisin. The Anglo-

Palestine bank, a major financial arm of the Zionist movement, was also involved in financing the scheme.[4] In turn, the financial and political backing of the Zionist Executive in London and influential Jewish British figures served as a highly useful bridge to key British policy makers.[5] Consequently, the grant of the concessions to Rutenberg met fierce public and diplomatic resistance from Arab leaders, who feared that the electrification scheme would allow the Jews to obtain "a stranglehold on the economic life of Palestine and Transjordania."[6] The construction of a powerhouse and the appearance of electric poles in Jaffa two years later served to corroborate these fears and triggered various public displays of Arab national sentiments. The suggestion that "Arab agitation" in general and recalcitrant Arab landowners in particular prevented Rutenberg from building an electric powerhouse on the Auja was therefore in line with the political reality of the times.

Nevertheless, this chapter finds that while Arab agitation was real, it mainly served as a discursive trope for justifying a decision prompted primarily by commercial considerations and the technological imperatives of machines. Arab agitation—neither in the form of recalcitrant landowners nor, more broadly, in the form of organized public and political opposition—did not determine the location of the powerhouse and the technology used for generating electrical energy. Neither did some presumed Zionist master plan. Rather, deliberations over location and the technology for producing electricity were an occasion for various ethno-national performances: asserting national concerns in newspaper articles and public speeches, associating electric light with matters of national pride or otherwise national humiliation, and scripting the relations between Jewish Tel Aviv and Arab Jaffa. The "concrete" process that solidified an attachment between electricity and ethno-national identities, in other words, should be looked at on a par with the "context" of relations between the colonial state, local governments, and the Arab and Jewish populations of Palestine.

But all this will come later. The powerhouse had just been completed, and the very process of its design and construction should give a first idea about the type of sociology that may be extracted from following the assembly of an electrical infrastructure.

A First Circuit: London, Jerusalem, and the Construction of a "Political Context"

Here, then, is this techno-political issue of Arab agitation. It is made of a set of connections whose materiality mainly consists of traces left behind by the circulation of requests and statements of appeal, confidential memoranda, government minutes, legal opinions, and various other textual forms. These documents flowed in the form of letters, hand-delivered notices, dispatches, and telegraphs mainly between Jerusalem and London, involving correspondence between Pinhas Rutenberg; High Commissioner Samuel and later High Commissioner Plumer; Secretary of State for the Colonies Winston Churchill and his successor in office, Leo Amery; Zionist officials and Jewish financiers; technical experts and electrical engineers; and a considerable number of officials in the Colonial Office and in the law offices of the Royal Courts of Justice.

Yet so-called Arab agitation was only one of many issues moving back and forth between London and Jerusalem. Among other things, there was the techno-political issue of Germany's electrical capabilities. While Arab-Jewish relations were mainly invoked at this point with respect to *where* to locate the powerhouse, British-German relations came to life over the question of who would assume a leading role in providing materials and technology for the electrification project.

The circulation of policy documents, legal proposals, and technical schemes regarding such issues were strongly personified in the international travels of Pinhas Rutenberg himself. Soon after being granted the concession, in September 1921, Rutenberg embarked on a trip from London to Jaffa with the intention of entering "into negotiations for the acquisition of land required for the Auja project" from Arab landowners.[7] He returned to London in late December 1921 and was soon off to Berlin to consult German electrical engineers and to negotiate the purchase of German machinery. On his return to London, sometime in mid-January, he received a letter, urgently hand-delivered by his brother Abraham,[8] from his lawyer and most intimate and faithful legal advisor, Harry Sacher. Having heard of Rutenberg's decision to build a diesel-fueled powerhouse, Sacher warned against the idea of dropping the plan to build a hydroelectric station on the Auja River. He minced no words and composed a carefully reasoned three-page letter, dividing his arguments into "financial" and "legal-political" aspects.

The financial aspect concerned the fact that Rutenberg wanted to purchase 150 dunam (roughly 42 acres) in the Auja, for which the Arab landowners asked £25[9] per dunam, bringing the estimated total cost to £3,750. Sacher concurred that the price was burdensome, yet reminded Rutenberg that they had also planned an alternative hydroelectric scheme that would have required the purchase of only 25 dunam. Besides, Sacher wrote, they had already spent considerable sums in preparing the plans, and "if the Auja scheme is dropped and fuel stations are substituted, then I would imagine that most of this money would be wasted."

Then came the legal-political considerations, which apparently were graver than the financial ones. "As you are well aware," wrote Sacher, "the Auja concession is for hydro-electricity and your right to erect [a] fuel plant is simply supplementary to your right to make use of the water of the Auja." In other words, "the setting up of a fuel station and the consequential abandonment of the hydro-electric scheme would, in effect, mean the dropping of the whole Auja Concession." Sacher concluded that Rutenberg would then have to negotiate a new concession and, horrified, he wrote that "as your legal adviser I cannot contemplate you with equanimity beginning all over again the negotiations." He warned Rutenberg that in all likelihood the British Government would not grant him a new concession because of the political circumstances that led to the granting of the original one: "The Government looks upon the Auja enterprise as a link to co-operation between the Jews and Arabs. If you were to drop it now, it would be a confession of failure which would have a deplorable appearance and effect."[10]

Sacher added another "political" consideration that proves important in tracing the connections that would lead to the fuel powerhouse. He reminded Rutenberg that the Municipality of (mostly Arab) Jaffa "approved of the Auja hydro-electric scheme" and cooperated in negotiating a supply contract. Dropping the scheme, especially if interpreted as a political failure, might drive the municipality to retract its approval. "A very grave blunder can now be made," Sacher concluded. "Your advisers in London and Berlin may have influenced you to reach a mistaken decision . . . and it is our duty to do everything to prevent it being made."

Yet, in spite of Sacher's dire warnings, Rutenberg was able to change course without losing the concession. In fact, by the time he received his lawyer's letter he was already busy placing orders for machinery from Germany and finalizing the plans for erecting a fuel power station. Moreover, while Sacher worried about a scenario that would link the technological change to the problematic political issue of Jewish-Arab relations, Rutenberg deployed the very same issue in order to pursue his new plan.

A string of documents and letters from the Electric Company, the first of which is dated July 24, 1922, and addressed to the Undersecretary of State for the Colonies in the Near East Department, provides an opportunity to trace the way in which the link between technology and politics furthered the move away from the hydroelectric plan.[11] This first letter does not announce the dropping of the hydroelectric scheme. It does not even ask for a postponement. Rather, it suggests a way to legally construct Rutenberg's actions as if work toward a hydroelectric scheme had actually begun: "I left for Palestine with the intention of acquiring land for the Auja Scheme," Rutenberg began, but "the prices demanded by the owners of the land were out of all proportion to the market prices." Article 14 of the concession authorized the Government, under certain conditions, to expropriate needed land and installations and to pay owners "fair" compensation to be financially borne by the concessionaire. However, Rutenberg explained that, for political reasons, "I agreed with the Government" that it was inadvisable to do so. Rather, "I decided to build a fuel power station in Jaffa," he wrote, with the expectation that work would begin in August 1922 and conclude by January 1923 (eight months prior to the date set in the concession for finalizing the hydroelectric station on the Auja River). Moreover, "in view of the fact that energy will thus be provided before the time stipulated in the concession, I respectfully beg to request you to consider the building of the fuel powerhouse as the commencement of work within the meaning of the concession and release me from the commitment to finish the hydro-electric station until September 1923." Not having to work under the pressure of time, Rutenberg concluded, "I shall be able to soon reach amicable arrangements with the owners of the land on the Auja."

The concerns of his lawyer notwithstanding, Rutenberg expressed no intention of waiving the Auja Concession when he decided to build the powerhouse. And his plan worked. No apparent "grave blunder" followed. The Office of the High Commissioner secured the consent of the Colonial Office and responded within two days, welding diesel and hydroelectric technologies through a new contractual understanding: Rutenberg would finish the fuel powerhouse by January 1923 and in exchange would not have to commence work on the hydroelectric station until September 1923.[12] From then on—for a number of years stretching well beyond the completion of the powerhouse and in tandem with its extensive and ever expanding electric grid—Rutenberg used each new request to extend the date for commencing the hydroelectric project as a medium of change, as a means to make slight yet meaningful modifications in the type of connections between the construction of the fuel powerhouse, the concession, and the suspended construction of the hydroelectric station.

In November 1922, Rutenberg asked for and secured an extension for finishing the powerhouse, moving the target date from January to June 1923. But this was a small matter compared with the need to maintain the undisturbed connection between the fuel-generated electricity that was about to be produced and the right of holding the concession. The request for extending the completion date was thus used as well as an occasion for the periodic maintenance of this connection: "In view of the difficulties encountered by me in obtaining from the Arabs the land necessary for the works, I have in agreement with the government begun the erection of a fuel station in Jaffa."[13] The matter of not yet building a hydroelectric station was therefore neutralized, set in the background of the more immediate task of finalizing the fuel station.

The powerhouse was completed on June 10, 1923. The ceremony was over. Tel Aviv, but not yet Jaffa, was seeing the light of its first wired streetlamps. A day later, Rutenberg wrote a letter to High Commissioner Samuel from his offices at "the Power House, Jaffa" to inform him of the event.[14] Although it was merely celebratory in appearance, and although it did not ask for anything, this letter is significant, here again triggering a change in the assembly of the relevant connections: in his letter, Rutenberg informed the High Commissioner that after successfully negotiating with the Municipality of Tel Aviv, "there are now prospects that the Municipality of Jaffa will enter into an agreement with us for the supply of energy for street lighting." At the same time, Rutenberg again invoked

"Arab agitation" against his concession and explained that it "compelled me to postpone the work on the Auja, which was my intention to commence on the conclusion of the Agreement with the Jaffa Municipality." It was therefore "advisable," he wrote, to postpone the conclusion of such an agreement so as to calm the atmosphere. New links were established, this time between concluding a contract with the Municipality of Jaffa and commencing the hydroelectric works. And a slight modification was inserted: at first, the quite straightforward reason for the (temporary) suspension of the Auja hydroelectric project related to the prices that "agitated," or simply greedy or otherwise recalcitrant, landowners asked for their property. In this latest letter, this specific reason was turned into a more overwhelming bundle: Arab agitation in general. Tracing the string of exchanges between Rutenberg and the British authorities shows how a "political context" was assembled and how, over time, it came to be understood as a valid explanation for the change in technology.

This series of slight and accumulating modifications took final shape in a letter written in July 1923, the origin of which and the identity of its author are not entirely clear. A handwritten note—"handed to Mr. Vernon, 13 July"—was scrawled at the top. The first paragraph stated what by now had become undisputed fact:

> Immediately after the grant of the Concession in October 1921, Mr. Rutenberg entered into negotiations for the acquisition of land required for the Auja Project. Demagogic agitaters [*sic*] incited Arab landowners to oppose the scheme.[15]

The document's mode of address—discussing Rutenberg and his actions in the third person—made it look as if it had been written by an official of the Mandatory Government of Palestine. Perhaps this was indeed the case and Rutenberg had received a copy. Perhaps it had been written by Rutenberg or his lawyer and handed to a government official who used it to plead the case with the Colonial Office. Perhaps it was telegrammed to London from an officer of the Jaffa Electric Company.

At any rate, the letter puts a kind of official seal on the move from one technology to another: Arab agitators "incited the Arab landowners to demand exorbitant prices for the land, making the transaction commercially prohibitive. To avoid undesirable complications connected with expropriation of the lands, the government in consequence authorized Mr. Rutenberg in accordance with the terms of the concession to proceed with the erection of the [powerhouse] which is now completed, supplying energy to the entire population." Having reiterated this chain of events, the document informed the Colonial Office that "the government considers it advisable not to press the company with regard to commencement of the Auja work."

Yet, as mentioned previously, the ongoing reference to Arab agitation came hand in hand with Rutenberg's repeated claim that the Municipality of Jaffa wished to take part in the electrification scheme and to enjoy its expected benefits. A subsequent directive from Rutenberg in London to company executives in Jaffa—written on July 31, 1923 (just two weeks after the previous document)—bore the title "Instructions to Jaffa Office": "Endeavour to purchase Hadrah and Jerishe [on the Auja] for an amount not exceeding 3000 Pounds. The land must include the Island, land around Hadrah to a level of six meters, and a sufficient area as is required for the work. Endeavour to be made to complete transaction by August 20. If impossible to buy both dams, try to buy Hadrah only for about 1500 Pounds."[16]

Thus, at the same time that Rutenberg was seeking relief from having to build a hydroelectric station on the Auja River (specifically in the Hadrah and Jerishe areas) because of Arab agitation, he was attempting to buy the very same land for these very same "exorbitant" prices. Later documents and correspondence clarify that Rutenberg wished to purchase the land on the Auja (with its dams and mills) in order not to build a hydroelectric power station but to use the water for irrigation. We are thus left to ponder whether "Arab agitation" may indeed adequately account for Rutenberg's shift from one technology to another and subsequently from one location to another.

What Did the British Know about the Auja River?

As previously explained, Arab agitation was not the only issue that circulated between London and Jerusalem. To move ahead, we must now turn away from the flow of legal and quasi-legal texts and examine more closely the flow of technical details concerning the electrification of the Jaffa region and the original Auja hydroelectric scheme.

The plan was to raise an "out-of-repair" small dam (Hadrah) by 1 meter (3 feet) so that the water level would reach 6.90 meters (21 feet) above sea level. Then the idea was to excavate a 2.6-kilometer (1.6-mile) canal that would create a "tail-race" of the flowing water at sea level. Allowing for loss, the "head" available for use in the water turbines would be 6.25 meters. At this point, the water would drive turbines and alternators installed at a nearby flour mill (Jerishe). Rutenberg and his team calculated that normal conditions in dry weather would generate a flow of 8.5 cubic meters per second and that this would suffice for producing 530 horsepower in the turbines (265 horsepower for each turbine plus one in reserve), or the equivalent of 350 kilowatts of electrical energy. From the hydroelectric power house, a 3-phase at 50 periods (cycles) of electricity would then be transmitted by 6,000-volt high-tension cables to the Jaffa District, supplying electricity for irrigation, water pumps, private and public lighting, and industrial enterprises. One of the high-tension lines would run to Tel Aviv, where it would be joined to an underground cable that would take the supply to Jaffa. Five transformer substations would be placed in Tel Aviv and Jaffa, where energy would be transformed into low pressure of "not less than 380 volts between phases and 220 volts to neutral."[17]

Historians who have considered the "political context" of granting the electricity concessions to Rutenberg have paid scant attention to the technical aspects of his scheme. Yet the quite established administrative routines in the Colonial Office with respect to authorization of concessions in the colonies were based on a three-phase coordinated process: principled political approval, negotiation of contractual terms and other legal preparations, and expert opinion on the technical and financial feasibility of proposed projects.

In matters of electricity concessions, the Colonial Office's most trusted experts were the electrical engineering consulting firm of Preece, Cardew & Rider,[18] which, directly communicating with the chief engineer of the Colonial Office, routinely commented on many colonial concessions in Palestine and elsewhere. In some cases, most notably Rutenberg's later proposals to electrify the railways, the firm engaged in lengthy discussions and made specific technical and financial recommendations.[19]

The principled political decision to promote Rutenberg's Auja scheme without first issuing a public tender was made in April 1921, when a draft agreement circulated among British officials in Jerusalem and London on the one hand and between Rutenberg and his team of lawyers on the other.[20] Colonial Office documents dated July 12 and July 16, 1921, instructed the department's lawyers to work out the legal aspects of the concession and to produce a workable draft "in the form used in Colonies and Protectorates."[21] Yet at least one official in the Colonial Office seemed to have noticed the administrative asymmetry between the political-legal phases and the technical phase of the process and expressed puzzlement over the haste: "Is not this concession a matter upon which we should obtain the opinion of Messrs. Preece, Cardew and Rider as we have done in other cases, for instance [the] Freetown [Sierra Leone] Electric Light concession? I am inclined to think that we should."[22] This communication went unanswered. On July 2, the Colonial Office informed Rutenberg that the Secretary was "prepared to authorize" the Auja Concession and that he could go ahead with his preparations.[23]

By the summer of 1921, then—just before the signing of the Auja River Concession on September 12, 1921—the considerable political and legal preparations for it had not been matched by a thorough technical examination. A letter from the consulting engineers to the Colonial Office on August 8, 1921, discretely mentioned that, in contrast to other concessions and agreements deliberated between Rutenberg and the Government, the Auja Concession had not been referred to them for examination.[24] This time a link was established: On August 24, the Colonial Office formally asked the consulting engineers to deliver a report on the Auja hydroelectric plan.

A week later, on August 31, twelve days before closure, the consulting engineers provided the Colonial Office with a seven-page report on the Auja plan.[25] The report stated that the engineers had met Rutenberg in person and that he had handed them four numbered plans: a general scheme (plan no. 302), a survey of the Auja River (plan no. 339), a sketch of dams (plan no. 306), and a power station plan (plan no. 307). The report first referred to the scheme's hydraulic aspects, emphasizing that Rutenberg presented them with technologies of "automatic discharging syphons and spillways for dealing with flood water" as well as with designs for "sluices and screens to control the flow of water to the power house." The report then commented on the electrical aspects. Referring to Rutenberg's estimates of the expected kilowatt demand, it noted that if demand exceeded 350 kilowatts, a diesel engine would be required as a supplement. It also stated that there was "nothing to criticize" in respect to the proposed distribution system and that the plan as a whole looked entirely "modern." All in all, the report concluded, there was "considerable evidence" of the thoroughness and care invested in Rutenberg's plan: "We are of the opinion that the proposed scheme is a practical one and that it has possibilities of being a sound commercial success."[26]

At the same time, the consulting engineers reflected on their own limitations, specifically implying a broken link between "expert opinion" and "local knowledge": "We have no knowledge of the district served," they stated and then: "We have however no local knowledge as to possible engineering difficulties which might arise and no means of checking Mr. Rutenberg's figures." Perhaps to compensate for the partial expert knowledge established regarding the concession, the report recommended that all electric works be carried out "in the mode and with materials or apparatus which comply with the Standards of the British Engineering Standard Association."[27] This recommendation was not adopted.[28] On September 1, 1921, Churchill sent a telegram to High Commissioner Samuel, informing him that the consulting engineers had produced a favorable report on the Auja scheme.[29]

The technical details of the Auja hydroelectric plan retained their consistency throughout the process. The Preece, Cardew & Rider report accurately recorded the specifications that Rutenberg provided in his proposal, and these very same specifications were copied, on the receipt of the consultants' favorable report, into articles 7 and 8 of the Auja Concession. So the Auja hydroelectric plan, up to this point, had been smoothly integrated into the first-order circuit that we have so far considered: technical documents joined the circulation that flowed between officials and lawyers in Jerusalem and London, incorporating consulting engineers and technical know-how in the process without transforming it in any significant way. Still, there is a slight stoppage here, a slight interruption of the circulation, that expresses hesitation on the issue of "local knowledge." And it is here, precisely at this slight point of disjuncture, that the first circuit may allow for the loose insertion of a second one.

A Second Circuit: Tel Aviv, Berlin, and the "Technological Context"

The second circuit is made up of materials produced precisely by that "local knowledge" that Preece, Cardew & Rider did not have at their disposal. It consisted of texts and exchanges and players that by and large circulated between Tel Aviv and Berlin. From early 1921 onward, the minutes of the Council of the Township of Tel Aviv were full of references to the need for wired electricity. By that time, Tel Aviv had some small 50-horsepower diesel-fueled generators and a modest direct-current distributive grid that supplied electric light and water-pumping power to specific locations and buildings.[30] Yet the township was ready and thirsty for greater electrical capacity and, accordingly, explored a number of options and proposals.[31]

Rutenberg simultaneously conducted negotiations with British officials in Jerusalem and London and with local officials in Tel Aviv throughout 1921. Enter April 4, 1921—five months before the signing of the concession—when members of the Council of Tel Aviv Township gathered for a routine meeting. Fifteen people were in attendance, and the meeting was chaired by Meir Dizengoff. Also present was Mr. Zeidner, who—reputably knowledgeable in matters of electric power—was about to "lecture" on Rutenberg's proposals:

> I saw the letter of Rutenberg to the Mayor of Jaffa in the matter of light and I also want to express my opinion about this whole business of light and water-supply in our township. Rutenberg plans to provide light and water-supply to Jaffa and Tel-Aviv and electric power of irrigation to Petah-Tikva. *Electric supply by means of the Yarkon [Auja] River will definitely not be sufficient* [emphasis added] in order to also produce power for industrial purposes. Only in winter, when Petah-Tikva can do without irrigation, it may be possible to receive electrical energy for a few small factories. Yet for us this is precisely the main issue, because both Jaffa and Tel Aviv are undergoing rapid development, and electrical power will be needed in order to establish many new

> factories. I am not saying that the Auja plan is unimportant . . . but from what we know from various European countries it will be possible to increase the energy if we install a "spitzen kraftwerk" [engine power plant]. . . . Maybe the company that seeks to exploit the Auja can do it for us, but perhaps we can also do it on our own. (Faris 1936: 280)

For people with "local knowledge" the opinion expressed above could not have been very surprising. After all, the Auja River was not very much of a river. In reality it came much closer to what others would have called a stream, or a brook, and it was practically common knowledge that the Jordan River was "the only considerable flow of water in the country."

A council member asked, "Can we get the consent of Rutenberg in case we decide to act on our own?" Another stated, "We must tell Rutenberg to install such a power-station for us and to include it in the 6000 Pounds that he wants from us for laying out an electric grid." Dizengoff intervened: "We need a special power-house for light and power, and we hereby request Mr. Zeidner to prepare a response to Rutenberg and to let him know that we are in great need for electricity."[32]

In a follow-up meeting, Dizengoff announced that "Rutenberg expressed a principled consent to build us a diesel powerhouse *before* he goes on to execute the Auja plan," and the council subsequently authorized the required expenditure.[33] A few months went by, in the course of which Rutenberg negotiated with the Council of Tel Aviv (and with the Municipality of Jaffa) the contractual details of the creation of an electric grid and the supply of electricity. The council members in Tel Aviv gathered for another meeting on February 5, 1922, when Rutenberg's concession for the Auja had already been publicly announced. The agenda included a report by Dizengoff about "the state of things" with respect to electricity, which was duly recorded in writing:

> The Chairperson lectures about the results of his trip abroad, an absolute necessity following the dire state of our electric installations and the urgent need for electric light, water-supply, and industry. Many industrial entrepreneurs arrived here over the past year and could not find their place in the absence of electric energy. Dr. Zlotsisti made a trip last summer on our behalf and entered into negotiations with the German firm AEG [Allgemeine Elektrizitats-Gesellschaft]. . . . AEG agreed to install for us a diesel-engine power station

> with the capacity of 350HP . . . and he telegraphed us immediately upon reaching this agreement. Yet we took into account that since then Rutenberg had acquired the concession and we stopped negotiating with AEG. I travelled with Rutenberg to Germany in order to see what needs to be done. We started in Berlin where we negotiated with Mayorchek, the chief engineer of AEG, about how to set up the installation. Mayorchek explained to Rutenberg that *the power which is needed is greater than that which may be generated by the Yarkon [Auja River]* [emphasis added] . . . and in consultation with AEG we concluded that we need to build a 1000HP engine-based power-house for Jaffa and Tel Aviv and *Rutenberg changed his opinion upon hearing Mayorchek and gave up on the idea of the Yarkon [Auja River]* [emphasis added] and so we decided to build a power-house for the District of Jaffa with 1000HP, and we already ordered two 500HP diesel engines that will arrive in five months' time . . . and then I went to the factory of Deutz in Köln [Cologne]) and realized that it would take 7–8 months for the engines to arrive.[34]

Both a local electrical engineering consultant in Tel Aviv and a German AEG engineer in Berlin had serious reservations about the potential of the Auja waters to produce sufficient electrical energy. In the Dizengoff report just quoted, we read that "*Rutenberg changed his opinion upon hearing Mayorchek* [emphasis added]." The original text was in Hebrew, and the word that Dizengoff used to describe this change of opinion goes further than the English translation allows. Dizengoff said that Rutenberg "*bitel da'ato*," literally meaning that he "dutifully cancelled his mind" or "obediently surrendered his opinion" on hearing the German engineer.

So, by January 1922, Rutenberg's lawyer had been informed of the course correction. Records in the company archive also marked January 13, 1922, as the date on which orders for machinery for a fuel station were placed in Berlin. And in February 1922 the Council of Tel Aviv was openly and candidly informed of the plan to build a diesel-fueled powerhouse for the township. The AEG engineers in Berlin also knew about it—in fact encouraged it—and only British officials in London seem to have been out of the loop until July 1922 (see Rutenberg's letter to the Colonial Office, discussed previously). Considering this circulation of "local knowledge" and looking at it from the perspective of local municipal figures, it becomes clear that Rutenberg's decision to build a diesel-fueled power station in Tel Aviv had stronger foundations than simply having to succumb to "Arab agitation." At best, we may assume that sometime around November or December 1921 Arab agitation on the Jerusalem–London circuit was acting as a "tail race" that created an energetic "head," accelerating the speed of circulation on the Tel Aviv–Berlin circuit.

So there are two separate circulations here, two circuits of knowledge, information, documents, and plans. There was a slight diversion in the flow on the first circuit, detectable in the technical report of Preece, Cardew & Rider, which somewhat paused while acknowledging a possible gap between expert and local knowledge. It was this diversion that opened up the possibility of looking at a second circuit, made up of other technical opinions, local discussions, nodes of negotiation and, most important, equipment and machinery required for the project.

Imperial Machines

Germany had been militarily defeated by England just four years earlier but, judging by the might of its electrical industries, it was nonetheless winning a war on the economic front.[35] "With good reason, then" writes Hughes, "[Berlin] has been called the Elektropolis. . . . Two of the world's leading manufacturers of electric machinery, Siemens & Halske and AEG, had central offices and factories in Berlin and its immediate surroundings" (1983: 177). By 1900, AEG was the largest manufacturer of electrical machinery in Germany, surpassing Siemens. It also financed, planned, and built power stations for electrical utilities, and designed and built electric streetcar systems. By 1903, writes Hughes, AEG had reached an agreement with General Electric in America "to divide their world markets" (1983: 179).

The company's decision to rely on AEG in Germany had been a sound one, not only because of the firm's quality products and technology transfer capacities but also because of its prices. In August 1921, before the concessions were granted and announced, Rutenberg was ready with a document bearing the title "Auja Scheme: Comparison of Different Tenders for Machinery and Materials."[36] The document was divided into seven chapters, each with a detailed comparison of British and German providers for one of the seven essentials on the scheme's shopping list: turbines, alternators, transformers, motors, poles (tubular), copper wires, and cables. The comparison included the prices offered by German firms such as AEG, Voith, and Siemens, and those of British firms such as the English Electric Company and Vickers Limited of London. A summary page compared the various instruments and machines (Figure 2).

	DESCRIPTION	NUM-BER	ENGLISH TENDERS			GERMAN TENDERS		
			NAME OF FIRM	£.-.-	%	NAME OF FIRM	£.-.-	%
1.	TURBINES	3	VICKERS LIMITED.	5350	184%	VOITH J.M.	2880	100%
2.	ALTERNATORS	3	" "	5661	197%	SIEMENS SCHU K.	2856	100%
3.	TRANSFORMERS	17	" "	2567	143%		1788	100%
4.	MOTORS.	94	" "	4800	150%		3200	100%
5.	POLES	150t	MANNESMANN CO	7500	214%		3500	100%
	Total...			25878	181%		14224	100%
	Sea transport to Jaffa	250t	£6. -per ton Charges Brit. port to Jaffa	1500		£2. -per ton Charges Hamburg to Jaffa	500	
	Total cost cif Jaffa			27378	186%		14724	100%
6.	COPPER WIRES cif Jaffa.	60t	BRITISH INSULATED & HELBSY CABLES LTD.	6360	108%	SIEMENS SCHUKERT BERLIN	5720	100%
7.	CABLE cif JAF.	3500 metr	" " " "	1225	146%	" "	836	100%

FIGURE 2 *Comparison of English and German machines and tenders*
SOURCE: ECA 2371-1-85.

When it came to hydroelectric turbines, the difference in price between German turbine manufacturer Voith and British manufacturer Vickers stood at 164 percent in favor of Voith.[37] British firms fared slightly better in the category of copper wires, showing only an 8 percent difference in favor of Siemens over British Insulated & Helsby Cables

Ltd. Moreover, the summary showed that the cost of shipping from Hamburg to Jaffa was considerably lower than the cost of shipment from British ports. With this, the circuit was firmly established, circulating equipment, know-how, experts, municipal officials, and entrepreneurs between Tel Aviv and Berlin. The initial phases of the material electrification of Palestine, in short, were in German hands for all practical purposes. And this second circuit, in turn, was only loosely tied to the Jerusalem–London first circuit.

The Auja Concession did not create a binding obligation on the concessionaire to purchase materials in England.[38] Yet Rutenberg was highly aware of the expectations of British officials that he would obtain the required materials there. After all, it was the British government—through its Colonial Office in London and the mandate authorities in Palestine—that had authorized and sanctioned his plan. Buying from British industries would be in line with Imperial interests in enhancing the demand for British exports and, by extension, relieving then prevalent unemployment at home. Anticipating potential political fallout in London and yet interested in what he perceived as superior German products and expertise, Rutenberg sought the approval of the Zionist Executive Committee, whose investments and shareholding constituted the financial backbone of the project.

Rutenberg wrote Dr. Chaim Weizmann, head of the Zionist Executive Committee in London and received assurance that he could proceed on the German track. A letter to Rutenberg from the Executive's office in Great Russell St., London, stated that "Dr. Weizmann handed to me your letter of July 21 and wishes me to say he is of opinion you should buy in the best market without consideration of any other factors."[39]

Rutenberg followed up by writing directly to Sir Winston Churchill, Secretary of State for the Colonies (Figure 3). In this letter, he sought to establish a connection between the second circuit, the circuit "on the ground," and the imperial-political materials through which the first circuit—the myriad requests, statements, and documents—had been sustained. This connection relied on the imperial reasoning that guided the British in their decision to grant the concessions to Rutenberg in the first place: promoting the Jewish National Home through the "future economic development" of Palestine. Rutenberg made links between the technical aspects of the electrification scheme, the lower prices of German equipment, and British colonial interests. First, he referred to the government's association of the concession with the prospects of electrifying the Jaffa–Jerusalem railway and possibly other railways throughout the Middle East. Higher electricity rates, Rutenberg argued, would raise the cost of transportation, thereby inhibiting economic development. Second, he explicitly reiterated the idea that a commercially successful electrical scheme would develop the country as a whole and therefore had the potential to pacify possible tensions between Arabs and Jews. The remarkable personal style of the letter and the bold request for an immediate face-to-face interview on Christmas eve is testimony to the unique bond that Rutenberg thought he had established with Churchill.[40]

21st December, 1921

The Rt. Hon. Winston S. Churchill, P.C. M.P.
Secretary of State for the Colonies, Colonial Office,
Downing Street, S.W. 1

Sir,

May I at the outset apologise for encroaching upon your valuable time, but the urgency of the following matter makes it imperative that I should address you with as little delay as possible.

In the spring of this year I collected data regarding the cost of the machinery and materials required for my project. I found that the lowest English tender was 98% in excess of tenders made in Germany, when the Mark was 1800-2000 to the £ sterling. Although German prices have since risen, yet the drop in the value of the Mark still maintains a difference of considerably more than 100%.

The purchase in England of the entire machinery and materials necessary for my project and for the elctrification [sic] of the Palestine Railways would accordingly involve an additional expenditure of several hundreds of thousands of pounds, and thus increase the rates of energy, rates of transport and cost of living in Palestine and so retard the economic development of the Country.

The placing of orders in Germany, however, might lead to undesirable political consequences in this country.

(*continued*)

These considerations I have submitted to Sir Herbert Samuel, who was of the opinion that the materials might have to be purchased in the cheapest market.

When I presented my plans to you in Jerusalem you were good enough to propound to me your sceme [sic] for the acquisition of machinery from Germany by way of reparations due to this country. As this matter has now reached a practical stage I am most desirous having a conversation with you on several vital points in this connection.

The beginning of practical work in Palestine seems to me to be of the greatest importance. The Arab Jewish difficulties can be readily solved by their participation in practical and reproductive work in this country. Further, in the Concession which your department thought fit to grant me on 12 Sept. 1921 it is stipulated that I am to form my Company and commence work before the expiration of 9 months, i.e. 12 June 1922.

I intend to spend the Christmas Holidays in making a personal study in Germany of the questions to which I have referred, but before I go, I consider it of the first importance that I should have your opinions and observations in the matter.

In view of the fact that I propose to leave for Germany tomorrow evening, you will appreciate my desiring such an interview as soon as possible.

I have the honour to be, Sir,
Your obedient servant,
(signed) Pinhas Rutenberg

FIGURE 3 *Letter from Pinhas Rutenberg to Winston Churchill, December 21, 1921*

A tight link was therefore forged between the technology that would be used for electrification and the country from which it would be acquired. In the summer of 1922, this link became news in London. Several historical accounts provide detailed descriptions of the response of British members of Parliament and the British press to what they saw as the abuse of a concession that should not have been granted in the first place. Most of these accounts focus on the fact that this change of course served as a convenient platform for "anti-Zionist/anti-Semite" political attacks on the close and sometimes intimate connections between the Colonial Office, Churchill, Palestine High Commissioner Herbert Samuel, and Pinhas Rutenberg on the one hand and, on the other, British colonial policy in Palestine. Responding to this critique, Barbara Smith

(1993: 122) reports on a memorandum from Churchill to the Colonial Office giving instructions to stop Rutenberg from buying equipment in Germany. Others suggest that Rutenberg reassured the Colonial Office that he would not place orders in Germany "until further notice" and that he would buy materials and equipment in England as long as the difference in prices did not exceed 10 percent.[41]

The price of British-tendered copper wires, which were required for laying out the electric grid, was only 8 percent higher than that of German wires. Neatly falling within the 10 percent difference, copper wires were indeed the single item on Rutenberg's initial shopping list that was bought in England.[42] It was only in the second half of 1923, when the powerhouse was already fully operational and the electric grid was undergoing rapid extensions, that Rutenberg instructed his managers to purchase two new 50-kilovolt-ampere transformers "possibly in England."[43] In late 1925, on the breakdown of one of the 500-horsepower Deutz engines,[44] the Electric Company for the first time bought two British Mirrlees diesel engines: a 250-horsepower engine and a 300-horsepower engine.[45] Shortly afterward, in order to further increase the capacity of the powerhouse, the company ordered from England an 1125-horsepower Fraser & Chalmers engine, which was expected to become operational in 1926.[46]

Thus, technology and machines and their origin played a material part in the development of the power grid. But what prompted the move from one electrical technology to another and, subsequently, from one location to another? The simple answer is that the Tel Aviv–Berlin circuit worked better than the Jerusalem–London circuit: under the best conditions, the hydroelectric project was designed to reach a maximum capacity of 350 horsepower at a time when anticipated demand required at least 1,000 horsepower. Considering anticipated local demand for electricity, available know-how, and affordable prices in Germany, it made sound commercial sense to electrify the District of Jaffa by means of a diesel-fueled powerhouse.

The more complex answer, however, must consider the heterogeneity of electrification. The Tel Aviv–Berlin circuit had no life of its own. The elements that circulated on it—machines, technical know-how, and agreements with local governments—had to be linked to the elements that circulated on the Jerusalem–London circuit in order to produce the novelty of electrification. The elements on the latter circuit—legal documents, technical and financial plans, and policy considerations—linked the Colonial Office, the Zionist Executive, the Jaffa Electric Company, and the Government of Palestine. It was on this circuit that the nature and future of Jewish-Arab relations were attached to the flow of electric current. On the one hand, Arab resistance to Jewish electrification was framed as a justification for dropping the hydroelectric scheme. On the other hand, the prospects of constructive cooperation between the Electric Company and the Arab Municipality of Jaffa were framed as a justification for the ongoing British support of a scheme that diverted from, if not flagrantly breached, the original obligations under the concession.

Arab-Jewish relations circulated on the London–Jerusalem circuit in ways that sometimes emphasized Arab hostility and sometimes highlighted Jewish-Arab cooperation, all in accordance with the concrete needs that this issue played in securing continued British support for the project. More than a reality that could be unambiguously ascertained, the Jewish-Arab question served as the basis for allowing the change in location of the powerhouse and especially the change to a cheaper and more applicable technology that flowed on the Tel Aviv–Berlin circuit. The Jewish-Arab question connected the London–Jerusalem circuit to German technological equipment and know-how, and it attached them both to local demand in Tel Aviv. The two circuits were connected, not merely by the fact that Rutenberg and his company were themselves dynamos circulating on both. And it was in the course of this movement that "Jews," and "Arabs," and the relations between them, were actively shaped, performed, and at least to some degree assembled.

2

Let the Poles Act

HUMANS AND MACHINERY, administrators and day laborers, lawyers and engineers, generators and application forms; these were some of the actors who took part in the assembly of the grid. The newly established Jaffa Electric Company (hereafter the Electric Company or the company), which was registered in May 1923, hired a photographer to follow the process from start to finish. In pictures of the arrival of electric generators, the shadows of the camera and the cameraman extend toward and join the machines, people, and horses whose power was needed to make the final passage from the Port of Jaffa to the awaiting sites (see Figure 4). Equipment was brought in either as complete assemblies or as components to be assembled on the ground. The first two 500-horsepower diesel engines were bought from Deutz in Köln. Other vital machinery such as generators, transformers, control switchboards, and numerous other components were ordered from Allgemeine Elektrizitats-Gesellschaft (AEG) in Berlin (Naor 2003).

Equipment was accompanied by the import of know-how. AEG acted as a consultant, and the Electric Company opened an office in Berlin. The AEG team in Tel Aviv consisted of six full-time consultants by mid-1923, headed by Mr. Herteuch. Rutenberg, spending most of his time in London and Berlin, issued instructions to his managers onsite: "Those in charge of the switchboard to follow the monitoring instructions

FIGURE 4 *New generators being transported to the powerhouse*
SOURCE: BNA CO 1069/731.

of AEG. . . . Herteuch must stay until the power-house is finished, then to be awarded with a wrist gold watch at the value of no more than 25 Pound Sterling with an inscription saying: 'in memory of participating in the construction of the first power-house in Tel Aviv'. . . . Stadi [another AEG expert] to oversee the wiring of Ajami [in Jaffa]."[1] In the District of Jaffa, where the Auja Concession legally applied, the first priority was to attract to the grid wealthy residents and interested merchants. However, the advance of wires and poles also depended on proper administrative and contractual connections with local authorities in Jaffa and Tel Aviv.

Regions, write Mol and Law (1994), do not reside in the order of things. They appear as distinct spaces when subjects and objects are clustered together and boundaries are drawn around them: difference becomes a matter of stressing similarities within each region and dissimilarities with others outside. Regions have to be produced, and one way of tracing their production is by looking at them as the products of networks.

Mol and Laws' study was of the disease of anemia, and they showed that different regions of anemia were the effect of a laboratory network for measuring hemoglobin levels. It is the ability to use the same unit of measure that enables comparative differences "across regions," and it is this insight into the capacity of networks to produce or at least enact the reality of different regions that guides this chapter. Jaffa and Tel Aviv were there "before" the electric grid, but I show that the grid took an active part in shaping the boundaries between them and in assigning each a different mode of attachment to electric current.

. . .

A handwritten letter dated December 27, 1925. It was in Hebrew, its style was meticulous, and its calligraphy was nothing short of beautiful. It was anonymously translated. The original writer was Fauzi Aboukadra, an Arab landowner from Jaffa, who politely addressed M. Dizengoff, the mayor of Tel Aviv. Aboukadra reminded Dizengoff of the latter's written promise of May 1923 to pay him the sum of 100 Egyptian pounds in exchange for his consent to allow a new street to be paved across his land, one that expanded the newly built commercial center between Jaffa and Tel Aviv. The street had already opened, wrote Aboukadra, but the promise was yet to be fulfilled: "Hence I beg to ask His Honor to inform me as to your willingness to pay or not." The letter had the latest date of those found in the Tel Aviv Historical Archive in a small bound file that contained earlier correspondence regarding this matter.[2]

The earliest letter in the file dated three years back to December 1922 and was handwritten in Arabic on Aboukadra's business letterhead. It asked Dizengoff to conclude the unfulfilled agreement. In its margins, a Jewish official of the Tel Aviv Township scribbled a comment in Hebrew: "The matter is urgent because Aboukadra sent a notarized warning to Rutenberg and threatens to sue and of course the township would also be embarrassingly implicated." In January 1923, Dizengoff replied to Aboukadra in French, apologizing for the delay and promising to address the matter within days. In March 1923, Aboukadra again inquired as to where things stood. Again written in French, the response informed him that his proposition to allow the development of a street on his land and the erection of electric poles along it in exchange for 100 Egyptian pounds, awaited the approval of the Committee of Merchants, which ran the affairs of the new commercial center. So not only Aboukadra was involved; the Township of Tel Aviv, and the Committee of Merchants but also Rutenberg's Electric Company, which had erected electric poles on the site were involved as well.

Aboukadra wrote another letter in September 1923. Yet another translation from Arabic to Hebrew, it shed light on the dispute. The letter opened with warm greetings for the Jewish New Year and proceeded to announce that the agreement between the parties was null and void because of the township's failure to pay consideration for the land, whose value, incidentally, had gone up 50 percent since the original deal had been made. Aboukadra further asked Dizengoff to inform Rutenberg that the electric poles on his land must be removed. Dizengoff, in turn, wrote the Electric Company, "strongly advising" it to pay Aboukadra the sum of 150 Egyptian pounds. Otherwise, "it will cost you 50 Pounds to remove the polls and re-place them on the other side of the street . . . because it was your responsibility to settle with the land owners before erecting the poles. . . . Split the cost with the merchants and pay Aboukadra, or else bear the unpleasant consequences of this business for both you and the merchants" (Dizengoff to JEC, 1923).

An unidentified company employee scribbled a response in the margins of the mayor's letter: "We only consented to arrange the wiring at your request, but it was the duty of the township to pay Aboukadra precisely in the same way that you previously paid him 50 Pounds for the private lot on which we placed a transformer."[3] So the laying out of an electric grid required wiring, but since wires could not just hang in the air, poles were also needed. And it was necessary to know where to position these poles, how far apart they should be, and how high. All of this required the preparation of maps and sketches and topographical measurements, as well as a considerable degree of local knowledge. Poles also had to be manufactured somewhere, so cost was a consideration.

Copper wires were imported from England, but poles had at first been brought from Germany.[4] They were heavy and cumbersome, and shipments by sea were slow and expensive. For these reasons, the Electric

Company soon invested in its own workshop for the manufacture of poles, and its first annual report stated that, while poles ordered from Germany cost 19 Egyptian lira (pound) per ton and locally manufactured poles cost 31 Egyptian lira per ton, the new workshop manufactured them for only 16 Egyptian lira per ton.[5] And then the company needed to position the poles, prepare maps and sketches, coordinate pole locations with municipal planning committees, gain approval from the Department of Public Works, and hire supervising engineers. Only then could the poles be placed in the ground. But who was to pay for this complicated endeavor? Residents and merchants who wished to have electricity wired into their houses and shops? Municipalities that sought to further the progression of public lighting? The Electric Company, which in due time would reap the profits from consumers?[6]

We will soon return to the matter of Aboukadra and pursue the tracing of the expanding electric grid. Here a slowdown is in order, clarifying the terms of reference. The electric grid is not only an assemblage of poles, wires, and electric current. It is also not just a product of political, administrative, or legal arrangements or just the outcome of certain social forces (e.g., ethnic, national) that may be isolated and treated as if they existed independently of the technical aspects of the material grid and hence could explain its assembly. The premise here is that the electric grid is a heterogeneous work of assembly that requires the development of standardized forms of exchange and established patterns of communication in the form of grid extension orders, serially numbered poles, stable units of measurement, map sharing, and other such means for calibrating technical imperatives with legal and administrative and politically "sensitive" demands.[7] I treat the electric grid as a social formation not only (or merely) in the sense that its assembly and expansion depend on various "social" matters (e.g., political context, organizational capacities, human labor) but in the sense that the flow of electric current through wires is a social movement in and of itself and not simply one that may be accounted for in terms of the social forces that push it or the social effects it produces: wired electricity connects heterogeneous elements, sets wired and unwired areas apart, allows comparisons between users, and provides different categories of consumption (e.g., public light, water supply, industrial demand, agricultural irrigation). The grid is therefore in the position to participate in shaping and defining distinct areas, or regions, of supply. This latter postulate will become clearer as the tracing of connections becomes more persistent. Therefore, it is time to return to the procrastinating case of Aboukadra.

Yekutiel Baharav, the general secretary of the Electric Company, addressed the matter of Aboukadra in a letter to the Township of Tel Aviv on October 7, 1923. This was a formal letter in both style and content, arguing that the township should not have approved the positioning of the poles if it had not previously settled the matter with private landowners. The same procedure had been followed when the township bought from Aboukadra the plot on which the company placed one of its transformer stations. Therefore, it was the township that had to bear the costs for removal of electric poles, the fault lying with "those who bought the street and did not pay." An unidentified hand at the receiving end scribbled a note on this letter as well: "What to do?"[8] Aboukadra's December 1925 letter, mentioned earlier, indicated that nothing was done in the following two years.

So poles had to be positioned and sometimes had to be removed. Once positioned, they occasionally triggered disputes and complaints. Aware of the power of poles and wires to cause damages to land and trees and therefore to bring about controversies, Rutenberg cabled specific instructions to his people on the ground: "Compensation for lands under poles and damages (gardens) must be fought out to minimum price. Assistance of government to be asked for. This extremely important with regard to future works. Eliminate by all means bringing these questions to court."[9] Avoiding litigation, the company settled in exchange for waivers: farmers and landowners signed forms confirming receipt of compensation for "trees cut down," "broken tiles," and the like (Figure 5).[10]

Some landowners did not easily submit to the obstructions caused by the presence of poles: "Satisfy most troublesome privately," Rutenberg ordered. "Give gifts such as lamps, fans, etc. but not money or free wiring."[11] Sometimes the Electric Company had to buy the land in order to erect poles and install its equipment. Such was the case, for example, when it had to position a transformer station on King George Boulevard in Jaffa and lay out wiring in the direction of the affluent Arab neighborhood of Ajami: "Instructions: buy land for transformer no. 5 and proceed

Abdel Magid Abou
Geeiten.

March 15, 1925,

I declare hereby having received from the Jaffa Electric Company Limited the sum of LE 1.500 (One Pound Egyptian, 500 m/m) for damages sustained through the erection of the High Tension Distribution Line to Sarafand, i.e.

For 4 Zanzilia trees cut down....	£ 1.200
For broken tiles................	0.300
	£ 1.500

and I further declare that I have no claims whatsoever against the Jaffa Electric Company Limited, and that I will not allow any trees on my ground to grow in the direction of the electric wires.

FIGURE 5 *Settlement agreement between Abdel Magid Abou, a landowner, and the Jaffa Electric Company, March 15, 1925*

with the erection of poles and laying of wires all over Ajami according to sketch no. JC20."[12] The company bought land, then, first in the District of Jaffa and soon afterward all over Palestine. One Mr. Itin was in charge of negotiating and concluding real estate transactions on behalf of the company. He received a commission for each successful deal and also ensured that each transaction was duly registered with the authorities.[13]

Yet sometimes poles did not trigger controversies and wires did not require settlements and purchases. To the contrary, it was sometimes the absence of a pole that was a cause of grievance, and it was sometimes the subsequent positioning of a pole that was a cause for relief to power-hungry residents. The presence of poles sometimes created a disturbance and sometimes a blessing. Sometimes landowners had to be compensated and sometimes not. Sometimes poles had to be removed and relocated, and sometimes their identity numbers and locations were duly affirmed. Sometimes the Electric Company had to purchase land, and sometimes it received municipal land for free. How, then, is one to make sense of the logic of the grid and the performance of the poles?

Thomas Hughes explains that the electric grid received its name because "the network of high-voltage transmission lines that would criss-cross the countryside took the shape of a great gridiron imposed upon a map of a country" (1983: 353).[14] An electric grid has the capacity to define a certain area by means of its "coverage" and thereby to also actively affirm some geo-social space (a region, a district, a country, etc.). At the same time, it has a remarkable urge to find new outlets for expansion, to locate new sources of connection, and to unite with neighboring grids. Electric grids are constantly on the move. They constantly establish new connections that may undermine the spatial fixity of any given area, and they have the capacity to transcend the boundaries (administrative, legal, or national) that mark areas or regions apart from each other.

The history of electrification contains many accounts of grids and regions. A study of electrification in the Madras Presidency of colonial India in the early twentieth century showed that linguistics-based politics that separated two regions of the Presidency—not just techno-geographical considerations—led to the creation of two separate electric grids (a hydroelectric grid in the South and a thermal one in the North) (Rao and Lourdusamy 2010). In Germany, a 1919 government plan to unite electric supply on a national basis—merging thermal and hydroelectric grids—failed to materialize because of the "particularism" and regional politics of private interests and local governments (Hughes 1983: 314–317). In the United States, early-twentieth-century plans for "superpower" and "giant power" systems, in which high-voltage transmission lines would integrate a plethora of local grids, triggered ideological conflicts between "state government and private enterprise" (Hughes 1983: 313).

The long-distance grid and the big central power station that supplied it were latecomers to the American countryside, where for many years isolated stations were the only form of available electric service.[15] Granovetter and McGuire (1998) show that there were several alternatives to central power stations and that isolated and small networks kept growing in size and number well into the twentieth century. In colonial India, opposition to the establishment of big power stations feeding a unified grid was based on Gandhian philosophy. Gandhi opposed large-scale electricity production and preferred equal distribution by means of small power stations in every village and town (Rao and Lourdusamy 2010: 62). And in the United Kingdom, legislation promoting the establishment of a unified grid in 1926 is hailed by Hughes as a significant accomplishment: "This achievement, in a country whose electric supply had long been parochial, was more political than technological and economic" (1983: 350).[16]

Almost everywhere, the initial spread of electricity had by and large been carried out on a town-by-town basis, where licensed or franchised companies contracted with municipalities for the provision of lighting and water supply. The development of small local systems of electrical distribution was uneven among regions, districts, and towns. Tinker (1968 [1954]) found that by 1937 only 143 out of 517 municipalities in India had access to electric service, the larger share being among bigger towns (191).[17] In Palestine, concessions for the production and distribution of electricity were granted to Rutenberg alone. An all-Palestine grid was envisioned, but it was one that would incrementally develop on the basis of contracts between the Electric Company and Palestine's towns and villages. In the District of Jaffa, the electricity supply was primarily oriented toward the electrification of the Municipality of Jaffa and neighboring Tel Aviv by means of a single grid that would be fed from a single powerhouse. And from the outset, the expansion of the grid toward these two places met different circumstances and imperatives.

Arab landowners were overwhelmingly more involved in cases in which the positioning of poles triggered negotiations over land values or disputes over damages. The spread of poles and wires encountered less friction when it concerned Jewish-owned land in Tel Aviv, where private landowners were mainly kept busy pleading to be connected and complaining about the high rates of installation and current. A site designated for a transformer station in Jaffa had to be acquired, while a site in Tel Aviv designated for same purpose was provided free of charge. In Jaffa, land had to be bought and landowners had to be compensated or otherwise settled with. In Tel Aviv, land was handed to the Electric Company for free.

A communication from an executive of the Electric Company to the Township of Tel Aviv in early 1923 indicated a possible source for this smoother circulation. Regardless of any other monetary arrangements, the executive requested that the necessary orders be issued to township officials "to assist us in all our needs, and authorize the free-of-charge use of streets and sidewalks for laying out the cables and the poles, and provide for the free-of-charge use of areas required for transformer stations as specified in red on the attached map."[18] Different flows were at work, one circulating through "the wires of Tel Aviv" and another circulating through "the wires of Jaffa." One connected the company's workers, equipment, and current to residents, businesses, and the Township of Tel Aviv; another, to those of the Municipality of Jaffa. Yet both worked through the same grid and shared the same power-generating source.

In contrast to other countries, where the challenge consisted of integrating separate grids that served different areas, the District of Jaffa experienced an opposite development. The process of electrification generated, on the same electric grid, a difference between two regions. First, it drew a physical boundary between Jaffa and Tel Aviv. Second, actual electrification was based on two distinct models of attachment to the wires: one that mainly reached consumers through the mediation of a local authority (in Tel Aviv) and one that mainly attached itself directly to individual consumers (in Jaffa). The electric grid, presumably blind and indifferent to ethno-national distinctions, was ultimately constitutive in affirming and drawing them.

One Grid, Two Regions

The process of both electrification and separation began in 1920, long before the granting of the concession and before the official marking of legal-administrative municipal borders between Jaffa and Tel Aviv. It commenced with the drawing of maps in Rutenberg's private engineering office in Jerusalem. These maps distinguished between two types of wires: low-tension wires for the lighting of streets and residential and commercial businesses were marked in blue; high-tension wires, the feeders of low-tension wires, were marked in red. Only one high-tension line, which ran roughly parallel to the Jaffa–Jerusalem railway tracks, was drawn.[19] It was these physical infrastructures—an existing railway line and a planned high-tension wire—that marked a border between Jaffa and Tel Aviv (Figure 6). No official border had been fixed at the time between the two towns.[20] It would also be wrong to assume strict ethnic

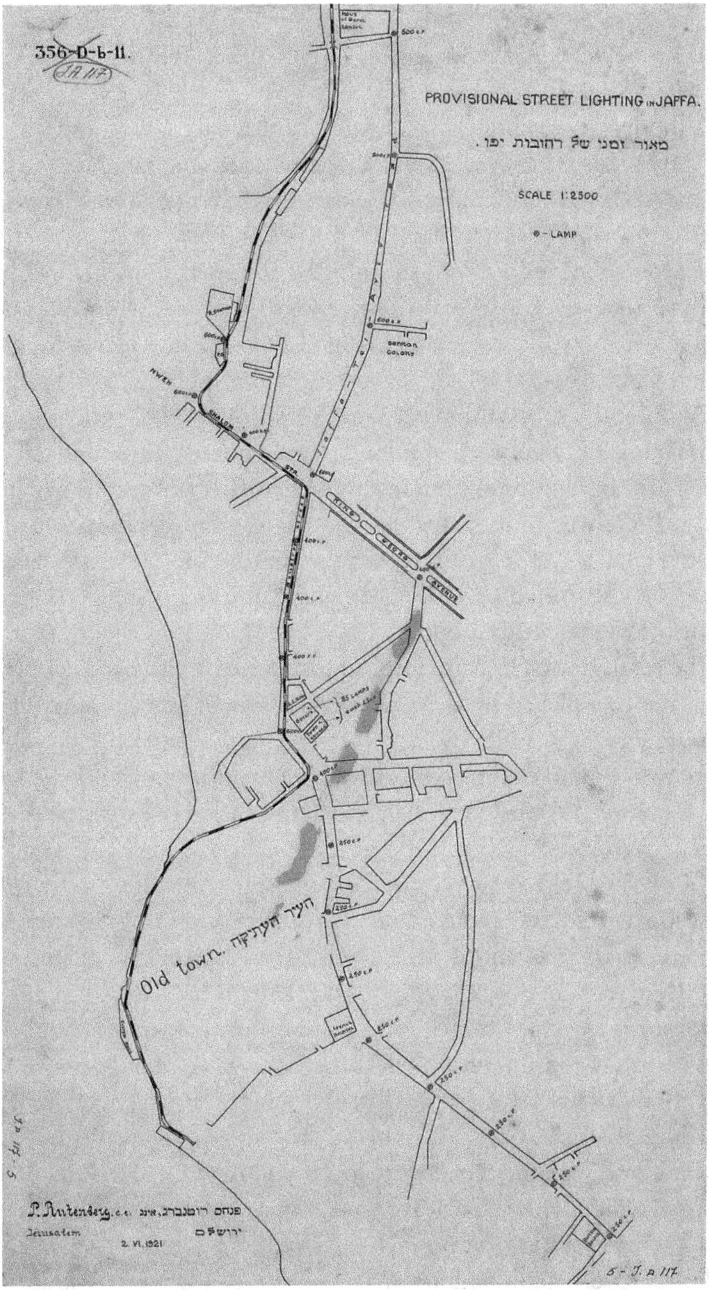

FIGURE 6 *Provisional street lighting in Jaffa*

segregation between Arabs in Jaffa and Jews in Tel Aviv at the time such maps were drawn.

Although it was bound to experience dramatic growth in years to come, early 1920s Tel Aviv was still a modest settlement compared to Jaffa. Of course, the intention here is not to suggest that there were no tensions between Arabs and Jews and that the distinction between the two ethnonational communities was not meaningful for local residents.[21] Rather, the point is that other markers of difference were also playing a meaningful part in everyday life at the time: religious distinctions (Jews, Muslims, and Christians), those between new immigrants and "natives of the land" (both Arabs and Jews),[22] those between "traditional" and "modern" communities, and those between employers and workers. Thousands of Jews lived in Jaffa, and almost everyone conducted their commercial affairs in that city. There were quite a few Jewish neighborhoods on the northern outskirts of Jaffa, but it would take years before they officially became part of Tel Aviv, sometimes only grudgingly and following complex negotiations (LeVine 2005, 2007).

The maps drawn in Rutenberg's office were significant not only because of the role played by the red high-tension line as a marker of two separate regions; the blue low-tension wires had a part to play as well. One such map marked low-tension wires extending in the direction of Manshie, a mixed Arab-Jewish neighborhood in Jaffa. Those wires came to what looked like an abrupt stop at the northern end of the neighborhood, bordering Tel Aviv, with no further visible connections. Two other low-tension wires that branched off the high-tension line exhibited the same pattern. The nascent township of Tel Aviv lay roughly there, where all of the low-tension lines ended. This was a map "for Jaffa," and one had to locate another map, "for Tel Aviv," in order to trace the route of those abruptly cut low-tension wires. The same pattern repeated itself for electric wiring maps for Tel Aviv. The low-tension wires also ended abruptly, as if leading nowhere. One had to bring the two maps together to see the full outline of the planned grid. The low-tension wires, still on paper, actively drew a distinction between two regions, however fuzzy it still was.

In fact, the erection of poles in the absence of a clear legally defined boundary between Jaffa and Tel Aviv sometimes triggered disputes and misunderstandings over issues of responsibility and jurisdiction. A letter from the mayor of Jaffa to the Electric Company dated March 1925 is shown in Figure 7.[23]

Municipality of Jaffa
No. 515
4-3-25.

The Jaffa Electric Company Limited.
Tel-Aviv

Sir.

The Municipal Engineer has drawn my attention to the line of poles which you have at present under erection in the Chelouche Street (separating Jaffa from Tel-Aviv).

These poles are being erected 2 meters out from the building frontage.

The footpath for this street according to the plan in this Office will be 1½ meter only and if the poles are going to be erected at 2 meters it means that they will be in the water channels and eventually cause considerable damage to municipal roads.

Under the circumstances I regret that these poles will have to be removed back ½ meter, at your earliest convenience. please. [sic]

Yours faithfully,
(sgd) A. El-Said
Mayor of Jaffa.

FIGURE 7 *Letter from A. El-Said, mayor of Jaffa, to the Jaffa Electric Company, March 4, 1925*

It was attached to a letter from the company to the Township of Tel Aviv: "We positioned the poles according to the instructions of your technical department, and we ask you to inform us under whose municipal authority this street lies, and if it is under your authority please take care of this matter with the Municipality of Jaffa."[24] Thus, the positioning of electric poles and copper wires on maps and their eventual location on the ground were active in drawing a border. This border did not obey, and certainly was not designed to convey, an ethno-national distinction between the two cities. The logic was commercial. The high-tension wire that drew a border between Jaffa and Tel Aviv followed a business plan for generating revenues from the sale of electricity: contracting municipalities and local authorities for water supply and streetlights and using the grid that would thereby develop in order to branch off to individual consumers. Simply put, the high-tension wire drew a border that presupposed separate contracts with Tel Aviv and Jaffa. This was a border that also corresponded to the then prevailing understanding of prospective buyers regarding electrical supply. The municipal officials in Jaffa and Tel Aviv who negotiated the future supply of electricity assigned to the material grid—poles, wires, and transformer stations—greater importance than that assigned to the intangible electric current it circulated. They treated poles and wires as "real" property that could be possessed in a way that electric current could not. The actual ownership of the material components of the grid was thus perceived as a guarantee that the municipalities would control the distribution of electricity within the municipal area. "I had a meeting at the Municipality of Jaffa today and they too are moving ahead," Dizengoff reported to the Council of Tel Aviv, and "in order to protect our citizens from losses and risks we firmly held to the principle that we have to receive the current en-bloc, in a monopolistic way, and it would be us who would distribute it among consumers . . . and the grid would be ours to keep."[25]

To the electrical engineers and experts who understood the technology, this must have seemed curious because wires and poles were good for nothing without a current to run through them, losing their value as property as soon as they were cut off from the generating source or when they failed to be technically maintained.[26] Still, the prospect of having wires that could carry electricity to any point that was adequately wired was as yet a novelty not entirely transparent to nonexperts (Faris 1936; Nye 1990). Even so, there was more to it than a confusion of form and content. Another connection could be made this way: the attachment of electricity to the ethno-national identity of its owners. The material possession of the grid served the Municipality of Jaffa in proving to opponents that, while Jewish Rutenberg held the concession for the production of electricity, the ultimate product would lose its "Zionist" identity; instead, control over the material grid was what secured the ethno-national identity of electric current. Thus, in return for its acceptance of the electrifying scheme, the Jaffa Municipality—just like Tel Aviv—demanded that the electric current produced by the company be carried to a "distributive center," at which point the municipality would take over: "The distributive center to be placed as near as possible to the town at a point fixed by the municipality. The distribution of power from that center to the actual consumers—houses, orange groves, factories—to be under the control and management of the municipality without any interference on the part of the company."[27]

The Municipality of Jaffa was not aware of the change in the electrification scheme until late 1921, and it assumed that electrical energy would be produced at the Auja (Yarkon) hydroelectric plant. On its withdrawal from the hydroelectric plan, the Electric Company built the diesel-fueled powerhouse precisely at the location that had been first imagined as the site of the future "distribution center." The powerhouse was built on grounds that the Council of Tel Aviv had bought from an Arab landowner and later transferred to the company without consideration.[28] However, this land was not part of Tel Aviv in an administrative sense, in line with an explicit plan "not to locate the power-house in Tel Aviv per-se but in a central place between Jaffa and Tel Aviv."[29] From a commercial point of view, as much as to address political sensitivities, it made sense to locate the powerhouse somewhere between Jaffa and Tel Aviv and to run a high-tension line roughly along the geographical interface between them.

It is precisely the existence of connectivity points between regions that highlight their network-enabled difference. These connectivity points may be detected on maps showing that Jaffa and Tel Aviv were indeed on the same grid, joined through a few points of connectivity where the marked poles on the maps showed a change from one set of serial numbers to another. Reading a map from the perspective of Tel Aviv shows these connectivity points as the wires and the numbered poles "ended" as they reached Jaffa. A similar reading from the perspective of Jaffa would have shown a mirror image of wires and numbered poles that "ended" as they reached Tel Aviv. A border was born, marked by a high-tension line, a limited number of connectivity points, and two subgrids, each with its own serially numbered poles.

Still, the grid would not have produced two regions if poles and wires had not acted in yet another way, one that produced a juridical-material difference between the circulation of electricity in Jaffa and its circulation in Tel Aviv. The Council of Tel Aviv took the logic of associating the value of electricity with the material grid to its ultimate conclusion: it contracted the Electric Company to build the grid "for Tel Aviv" at the town's own expense.[30] Strongly associating "electric value" with poles and wires—and convinced by the company to invest its own resources in building a grid that the company would have had to build anyway in order to generate revenue—the council sealed the deal by insisting that the company calculate the value of the town's practically useless direct-current electrical machinery and hardware as part of the transaction.[31]

From the perspective of the Electric Company, this type of contractual connection represented sound commercial prospects: a secured municipal area where the cost of laying out the grid would be paid for, with the further contractual obligation to pay additional sums for every extension that the municipality would ask for in the future, beyond the agreed on baseline. However, the contract also ensured that the company would not be prevented from laying out wires and positioning poles at its own expense wherever and whenever it identified commercial benefits, even if it were not asked to do so by the local authority.

And here lay the fundamental asymmetry between Jaffa and Tel Aviv. Municipal officials in Jaffa expressed support and even enthusiasm for electrification and had been in close touch with the Electric Company since early 1921.[32] Yet the municipality was also exposed to pressures to boycott "Jewish electricity." Several Arab newspapers, most notably *Filistin* in Jaffa, published commentaries and articles against Rutenberg's scheme.[33] The Jaffa branch of the Moslem-Christian Association also voiced opposition in the form of letters, pamphlets, and demonstrations. When electric poles made their first appearance on the streets of Jaffa, they set into motion sentiments that ran under the banner "The Poles of Rutenberg are Our Gallows."[34] Thus, the erection of poles in Jaffa triggered the performance of national sentiments. And as much as the prices asked by Arab landowners in the Auja basin had probably expedited the Electric Company's decision to change location and technology, so did boycotting pressures slow the determination of the Jaffa Municipality to conclude a contract with it.[35]

However, the absence of a contract with Jaffa was also a cause for complaint. Some Arab columnists criticized the calls to boycott electricity. "We are not acquainted with the cause that makes them oppose this scheme," one writer observed. "Is obscurity better for us than light? Until when shall we be sunk in the obscurity of generations and dark hateful intolerance? . . . The land has enough misfortunes, sufferings and obstacles in the way of its progress and prosperity." Ahmed Kamal, another writer, wrote that "every stone you throw at this scheme is an obstacle which impedes our advance to improvement and progress."[36] A group of Jaffa merchants were already complaining in July 1923, barely a month after the Jaffa powerhouse became operational. Writing on behalf of the Jaffa Chamber of Commerce, the group submitted a petition:

> The fact that the Municipality of Jaffa has refused to illuminate the streets of the main business section of the City with electricity from . . . Rutenberg's power plant in the time when various smaller and less important parts of the town are already illuminated, forced us to express hereby our deep protest against this offending resolution. We are dutiful tax payers, residents of Bustrus Street, King George Avenue, Mustakim, Station Road, and Suk El-Deir. Our shops are the heart of the business centre of the town and while other parts are already illuminated we do not get light because the Municipality still refuses a contract. We protest the reign of darkness in our streets. . . . Modern men suffer from the absence of light. . . . We need illumination as much as humans need their eye sight.[37]

In Tel Aviv, connections to private consumers were mediated through the township because it was the latter that determined the pace and direction of the grid by issuing "extension orders." In Jaffa, the grid extended on its own initiative. In November 1923, the Jaffa Municipality entered into a contract with the Electric Company only for the supply of street lighting. By then, the company had already wired those parts of the city that it considered to have potential for private demand. But this arrangement meant that it was on its own in terms of both having to secure individual contracts and having to negotiate the rights of passage for its poles and wires.

The Jaffa Municipality did not obstruct or interfere with the erection of poles and the laying of wires that went "over its head." The traces left behind suggest that the Electric Company could rely on the cooperation and active assistance of the municipality and its engineering and planning officials. "Obtain approval [of] Jaffa Planning Commission" for the location of transformer stations, Rutenberg advised company technicians, leaving behind the trace of a procedural-administrative connection between the municipality and the company. Publish advertisements for electric wiring in Jaffa Arab newspapers, including *Filistin*, he further advised, and "if necessary demand assistance from the Municipality."[38] Rutenberg was explicit on this issue, stating that "the Jaffa municipality has never expressed any opposition either to the scheme as a whole or to the erection of poles. On the contrary, the Mayor had on several occasions commented in the name of the Municipality on the advantages which would accrue."[39]

The overall effect of the maps, contracts, and materials that were needed for the assembly of the grid was a different model of grid expansion in Jaffa, one whose cost was borne by the Electric Company, relying on direct connections to domestic and commercial consumers without the full mediation of the local municipal authority (save for technical aspects). While the grid in Tel Aviv expanded through "extension orders" issued by the local authority, no similar pattern existed in Jaffa, where the pace and direction of the grid became a company affair in its relations with individual consumers.

The creation of a difference between those who are connected and those who are not is an elementary social effect of electrification. But the *type* of attachment between electric current and the single consumer also makes a difference. In Jaffa, (mostly) Arab individual consumers directly interacted with the grid. In Tel Aviv, the access of (overwhelmingly Jewish) residents to wired electricity depended on the substantial involvement of the township. Special measures had to be taken, educating the grid in how to expand in the Arab region and how to nurture relations with Jaffa consumers and officials. Rutenberg gave instructions in the form of a manual:

1. Buy land for transformer-station and obtain approval [of] Jaffa Town Planning Commission for location of transformer-stations.
2. Start as soon as possible the laying of cable . . . so that new parts of erected line be immediately included in the net. . . . Proceed with the erection of poles and laying of wires covering all Ajami. . . . Proceed with the work as

speedily as possible but eliminating unnecessary expenses.
3. Run advertisements in Jaffa Arab papers for two months and if necessary demand assistance from the Municipality.
4. Give special treatment to individual Arab applicants:
[A] At least first 100 applications to be received in this office, dealing personally in the most kind and friendly manner.
[B] Persons coming in touch with Arabs on house connections . . . to speak Arabic and treat Arab consumers and population in general in the most correct and friendly manner.
5. Change individual electric-supply application forms: Change vertical column first page: Arabic in the first place, Hebrew in the second.
6. Arab labor to be employed as much as possible.[40]

In both towns, being on the same grid, it was "strictly business." Electric company records consistently showed an uncompromising approach to questions of expenditures and prospective revenues.[41] As part of the company's business plan, a high-tension line drew a border, and a contract with the Township of Tel Aviv ensured that it would have its "own" grid on the general grid. The grid attached differently to consumers in Jaffa and consumers in Tel Aviv. An electric grid was assembled in the District of Jaffa, and the company that built and managed it was named the Jaffa Electric Company. Yet it generated two electrical regions and in the process enacted a divide between Arabs and Jews. We end up with one grid and two regions; one electric grid and two ethno-national groups that it performed and sustained.

The Tramway That Never Was

And then there is the story of a tram that never was: the plan for a public transportation system that had the potential to transcend the evolving boundary between Jaffa and Tel Aviv.

It was 1924, and preparations for the construction of an electric tramway were in full swing. Watchful observers collected descriptive statistics on the circulation of "vehicular traffic" in and between Jaffa and Tel Aviv. They used meticulous forms, distinguishing among "cabs, automobiles, and autobuses," arranging their counts on the basis of different days of the week. The data for one such week in 1924 showed that traffic peaked at 3,904 people traveling to Jaffa on November 16 and reached

	Tel-Aviv - Jaffa				Jaffa - Tel-Aviv.		
Date	Kind of vehicle	Number of persons	Total	Date	Kind of vehicle	Number of persons	Total
November 1924				November 1924			
16th	Cabs	249		16th	Cabs	342	
	Automobiles	280			Automobiles	329:	
	Autobuses	3375	3904		Autobuses	3147	3818
17th	Cabs	210		17th	Cabs	313	
	Automobiles	236			Automobiles	180	
	Autobuses	3066	3512		Autobuses	2764	3257
18th	Cabs	195:		18th	Cabs	285	
	Automobiles	211			Automobiles	195	
	Autobuses	3225	3651		Autobuses	2998	3378
19th	Cabs	216		19th	Cabs	243	
	Automobiles	220			Automobiles	189	
	Autobuses	3185	3621		Autobuses	2360	2792
20th	Cabs	266		20th	Cabs	254	
	Automobiles	175			Automobiles	230	
	Autobuses	3225	3566		Autobuses	2971	3455
21th	Cabs	184		20th	Cabs	236	
	Automobiles	141			Automobiles	160	
	Autobuses	2636	2961:		Autobuses	2279	2675

FIGURE 8 *Comparison of vehicular traffic between Jaffa and Tel Aviv*

its lowest point of only 2,961 travelers on November 21 (Figure 8). Buses were identified as carrying most of the travelers, thereby representing the main competitor to a future tram.

Many more people probably circulated the city on foot or by bicycle and cart, but no such statistics were presented. Observers also counted traffic between Tel Aviv and the nearby (and already electrified) Jewish agricultural settlement of Petah-Tikva, enlarging the potential arena of development. The count of vehicular traffic was eventually presented to the Electric Company to show that sufficient circulation existed to justify an investment in an urban tramway system.[42]

The company was approached by several interested parties who had prepared detailed proposals. One group consisted of local Jewish merchants from Tel Aviv who relied on the technical know-how of Siemens in Germany. They proposed to raise 50 percent of the necessary capital and expected the Electric Company to invest an equal share. Another proposal came from a group of Jewish industrialists in Egypt,[43] who proposed to rely on Swiss and French technical know-how. Their financial plan was to establish a joint company in which the Jaffa Electric Company would hold 51 percent of the shares. A final proposal came from businesspeople headed by Dizengoff (a businessman as well as the mayor of Tel Aviv), hoping to raise money from American investors.[44]

The more ambitious plan was to construct five urban, three suburban, and three interurban lines in the Jaffa District, altogether amounting to a capital investment of £250,000. A more modest plan focused on the construction of urban lines only, with a projected capital investment of £100,000. In all cases, the Electric Company was expected to invest in the projects, and, in all cases, the various entrepreneurs wanted to know what the company would charge for a kilowatt-hour. Again, in all cases, the written proposals and the meetings that were taking place with representatives of the Electric Company conveyed a sense of urgency.

This was not the first time that streetcars had been envisioned for Jaffa. Plans for horse-carried streetcars and light steam trains to nearby agricultural settlements had been proposed in the 1890s, shortly after the inauguration of the Jaffa–Jerusalem railway by Ottoman authorities (Cotterell 1989). At the time, such plans were considered uneconomical, but things had substantially changed by the early 1920s, and plans for streetcars resurfaced with the growth of the population and the prospects of electric power.

The urgency and seriousness that accompanied the various proposals seem to have been inspired by developments elsewhere. In Europe and the United States, the process of electrification was tightly bound up with the introduction of electric streetcars. Electric companies directly invested in the electrification of streetcar networks and relied on the supply of electricity as a considerable portion of their revenues.[45] Nye reports that the United States saw an "explosive growth" in streetcars around the turn of century, with General Electric and Westinghouse leading the industry (1990: 91–92). The rapid development of electrically powered networks of urban mass transportation—typically mediated and worked out through "charters" with local governments—provided the grid with a sense of direction and a rapid ability to expand. The same held true in countries neighboring Palestine. All of the major electric companies in Syria, Lebanon, and Egypt had invested in and relied on the supply of electricity to urban tramways (Faris 1936).

Nye (1990), offering a phenomenological analysis of electric streetcars in American cities, noted not only that streetcar networks reshaped the demographics of cities and the relationship between urban centers and suburbs, but also that they "created a new experience of urban space," new types of interactions among strangers, new forms of leisure and pleasure, and new opportunities for shopping and tourism (1990: 104).[46] Seemingly proposed under the influence of new experiences abroad, entrepreneurs in the Jaffa region urged the Electric Company to be aware of "the absolute necessity" of the project and warned that it could no longer be postponed. One such detailed proposal, written in French, explained why the project held out the promise of profits: no other region in the whole of Palestine, it stated, had been developing as rapidly and vigorously as the District of Jaffa. The urban core was already surrounded by agricultural settlements hungry to export their products through the Jaffa Port. Moreover, Tel Aviv had witnessed tremendous growth: from a few hundred residents by the end of the war to 20,000 people by 1924. The construction of a tramway would also further the development of industry, and this in turn would benefit the Electric Company. The Tel Aviv coastline promised a future rise in tourism, all in all creating greater demand for efficient urban transportation and hence greater demand for electric traction.[47]

Meir Dizengoff did not mince words: "We urgently need your response," he wrote to the Electric Company, "we are very interested in having a tramway as soon as possible because the traffic in the city becomes impossible by the day. . . . It is necessary for the development of the city and in order to prevent skyrocketing prices of land in its center. . . . Towns and cities all over the world enjoy tramways so please do not be too harsh in your terms." And he issued an implicit warning: "In case we shall not be able to reach an agreement on a joint tramway project we shall perhaps resolve to establish a bus-based system of transportation and then the participation of the Electric Company would not be necessary because no electric current would be required."[48]

There are few traces left of the company's response to this flurry of ideas and proposals. Some handwritten notes and comments suggest that the company repeatedly stated its exclusive rights to supply electricity to the projected tramway, but stalled on providing concrete answers with respect to its readiness to invest in it and the rates it would charge per kilowatt-hour. In contrast to what historians observe about the symbiotic relations between the development of electric grids and the spread of urban tramways elsewhere, and in contrast to the typical situation in which electric companies were leaders in pursuing the electrification of tramways, nothing of the sort took place in Palestine. Proposals were put on the table throughout 1924 and 1925, meetings took place, and then there was silence. All came to an abrupt stop—no tramway system was ever constructed in Jaffa and Tel Aviv. Perhaps the Electric Company was reluctant to invest in the project because by the mid-1920s there were already some reports of falling profits for electric streetcar ventures in the United States (Nye 1990: 134).

Or perhaps the project stalled because local indicators showed an economic slowdown, particularly in the Tel Aviv construction sector. However, this last possibility is highly questionable. The "economic crisis" arrived, or was at least acknowledged, only in 1925 and at any rate did not affect the soaring revenues of the Electric Company (Giladi 1971). Moreover, at roughly the same time the company began negotiations with the Government of Palestine for the construction of an electric funicular railway line on Mount Carmel in Haifa.[49] Finally, perhaps investors backed off. At any rate, it is certain that by 1925 the Electric Company showed no appetite for the project, particularly since it had discovered other lucrative outlets for the flow of current.

Writing about urban tramways in the United States, Kline describes the way they were "woven into the fabric of the city" (1990: 117), bringing closer together separate areas and distant neighborhoods. Analyzing the Transmilenio project in Bogotá, Valderrama shows how this closed-circuit transportation system resulted in the "physical reorganization" of urban public space, the reconfiguration of "a whole set of relations" within it, and the shaping of the identity and lived experiences of its people (2010: 133–137).[50] In Jaffa and Tel Aviv, such a potentially important urban unifier did not materialize, arguably deepening or at least failing to slow down the incrementally growing divide between areas for Arabs and areas for Jews.

The Power of Collective Extensions

The trajectory of the grid in Jaffa would not be forgotten. In years to come, its extensions through contracts with municipalities became quite standard all over Palestine. Yet the Electric Company preserved the principle that it could wire people even in places where municipalities or other public authorities were reluctant to contract it en bloc or lacked the resources to invest in electrification. When the Government of Palestine prepared a draft Electricity Ordinance, the company strongly objected to a requirement that it obtain the consent of each municipality before laying out a grid. It insisted that the ordinance should take stronger account of "local conditions": "It cannot take the form of an existing law in a cultured and developed country. . . . Under the present circumstances [this proposition] will be abused for political agitation, and create great difficulties both to the government and to the concessionaire."[51] The rhetorical attachment of electricity to ethno-national considerations preserved the Electric Company's ability to develop distinct regions on the grid, relying on direct connections to consumers in the absence of a cooperative local government.[52]

It was on the basis of connecting individual consumers that the grid expanded in the Municipality of Jaffa and later in Arab-Jewish Haifa. But the high costs of individual connections excluded consumers in smaller localities where residents could not rely on their municipality to invest in grid construction. The result was similar to what had been noted in the United States at the beginning of the twentieth century. Dominated by private industry and tied to prospects of profit, American long-distance wires assumed the shape of a snake, servicing electricity to "pockets of lucrative" business while bypassing unattractive townships and farms (Nye 1990: 303). Kline, studying the electrification of rural America, shows that "when electrical companies turned their attention to the domestic market in the 1920s, they focused on the densely populated and wealthier urban centers [and] were reluctant to serve farmers because of the high cost of building lines into sparsely populated areas not known for high disposable incomes" (2000: 132). The result of this pattern of connections was that "by 1930 electrification had decidedly increased the gap between the farm and the city" (Nye 1990: 303; also see Platt 1991).

While the United States experienced a growing divide between urban and rural connections, Palestine saw a significant difference between Jewish and Arab settlements. The first annual report of the Electric Company's board stated that a high-tension line to the British military installations in Sarafand was "almost completed and is awaiting only the government test." It explained that the high-tension line to military installations was located "conveniently near" some Jewish agricultural settlements and that negotiations for the supply of electricity to these places were under way.[53] The report of the directors for the following year (1923–1924) stated that the company had completed a 13.7-kilometer high-tension line to the army installations and that another 4.7 kilometers "branched off" from this line "to a Jewish agricultural settlement [Rishon Le-Zion]."[54] It did not mention any connections to Arab locations, in spite of the proximity of the same high-tension line to numerous villages and to the relatively big towns of Ramleh and Lydda. Maps drawn by the company by late 1926 also confirmed that Arab locations remained unconnected.

A similar pattern repeated itself all over Palestine in the years to come: extensive electrical connections to Jewish localities and meager connections to Arab ones. By 1926, only Jaffa and Haifa (both having significant Jewish populations) were on the grid, taking electric supply for streetlights and by and large facilitating connections to individual consumers. More than 20 localities were considered to be part of the electric market in the Jaffa District by 1931, but only two of them were Arab. Arab villages were not connected at all, yet Jewish settlements with populations under 2,000 increasingly attached themselves.

The growing difference between Jewish and Arab settlements in the number of connections to the grid provides another opportunity to examine the "political" explanation of electrification in Palestine: the possibility that this difference had to do with the Zionist character of the Electric Company (see Smith 1993), with Arab resistance to "Jewish electricity," or with both. Such explanations, grounding electrification in considerations of race and ethnicity, are offered by Chikowero (2007), who studied colonial Rhodesia (Zimbabwe), finding that electrification there developed as a racially based project: wired electricity was intended as a luxury good for white settlers and as an instrument of control over the nighttime movements of black natives. Beyond streetlights, wired electricity was supplied only to white residences in white neighborhoods, with the result that "electrification in Rhodesia did not simply exclude; rather, it played a crucial role in the creation of the 'slum'" (2007: 295). There is no evidence that a similar process took hold in Palestine.

The Electric Company invested special efforts in connecting Jaffa to the grid, in spite of municipal hesitations, and it acted in a similar way wherever it identified a potential for profit. To remove unwarranted obstacles, Rutenberg even required his clerks and agents to attend classes in Arabic that the company organized.[55] There is also very little evidence to show that the difference between the number of Jewish and Arab settlements that had been wired had to do with Arab resistance to Jewish electricity. The issue of Arab resistance was sometimes put on the table, yet it was not necessarily raised by Arab "agitators." Such, for instance, was the case in Haifa, where the Electric Company invested considerable efforts because of the area's rapid industrialization.

The circumstances surrounding the construction of the Haifa powerhouse in 1925 bore some resemblance to the construction of the Jaffa powerhouse two years earlier. From a formal legal standpoint, powerhouse construction in northern Palestine should have been authorized under the Jordan Concession, the terms of which stated that a powerhouse could be built as a backup diesel-fueled station for the planned hydroelectric works on the Jordan River. However, negotiations over the Jordan Concession were concluded much later, in March 1926, when the Haifa powerhouse had been operational for more than a year. Both the Electric Company and the government were aware that the electric grid was expanding faster than the legal arrangements that were supposed to authorize it. Both raised this issue in correspondence with the Colonial Office in an effort to expedite and conclude negotiations over the Jordan hydroelectric concession.[56] Here as well, the development of the grid was linked to "Arab agitation." This time, it was not the company but High Commissioner Samuel who raised the issue. He warned the Colonial Office that the Electric Company was "proceeding with the work of fuel power stations both at Haifa and at Tiberias and there is at present no legal authority sanctioning the actions of the Company in breaking up streets and laying wires, etc. . . . I apprehend that, unless early legislative authority is forthcoming, the acts of the Company may be challenged in the courts and injunctions obtained. Arab opinion is growing restive and protests are appearing in the press." Summarizing his concerns, Samuel wrote, "I cannot too strongly urge the desirability of definitely confirming the concession and validat[ing] it by the necessary legislation as early as possible."[57]

Yet again, and not unlike the case in Jaffa a couple of years earlier, the actual weight of Arab agitation and "political difficulties" was not very substantive on the ground. Quite a few newspaper articles appeared in the Arab press ridiculing the ignorance of those who objected to Rutenberg and complaining about unreasonable political impediments to progress.[58] Rutenberg, reporting to his shareholders in early 1926, announced, "The opposition of the Arab population which was encountered by the Corporation as a result of political agitation no longer exists and the Haifa Municipality [has been] a consumer of electric energy for street lighting since the middle of December 1925."[59]

It seems that the reference to Arab agitation had less to do with real obstructions to the expanding grid and more to do with the wish to accelerate the pace of decision making on the Jerusalem–London circuit and to calibrate the grid's material and legal elements. Basim Faris—an economist studying electricity in Syria and Palestine in the 1930s—reported that Acre was "served by the concession," although it had a reputation of being "a strong pro-Arab town" (1936: 61). Ramleh, initially ignored by the Sarafand high-tension line, was also connected toward the end of the 1920s. Faris had no doubt: "Economic demands triumph over nationalism" (1937: 67).[60] Yet he also found that Lydda, "which is ten minutes' walk from Ramleh, is still averse to such a convenience as electric current, and so is not as yet served; perhaps the low standard of living of the poor population prevents the use of the service at the present rates, which cannot compete with petroleum for lighting" (1937: 66).

Faris offered two complementary explanations for the unwired state of Arab settlements, one cultural and one economic. The cultural explanation corresponds to the idea that electrification requires "changes in mentality" (Coquery-Vidrovitch 2003: 347) and that "every new technology is a social construction and the terms of its adoption are culturally determined" (Nye 1990: 381). There is evidence to show that electricity was perceived as a luxury good for the urban rich both in the United States (Kline 2000) and in Palestine (Faris 1936) (and roughly at the same time). Consequently, writes Kline, wired electric supply reached rural areas accompanied by a variety of educators (e.g., home economists and salespersons) who took it upon themselves to change and shape new consumer habits and new understandings of modern life. Kline also notes the publication of training brochures to equip farmers with new knowledge about the merits of electricity, aiming to convince prospective consumers that electricity was preferable to kerosene and coal for light and heating (2000: 150).

Faris's work represents a similar approach; the purpose of his book, he explained, was "to acquaint the peoples of Syria and Palestine with the actual conditions of the electric industry, enabling them better to appreciate and evaluate the problems arising therein." Thus, he stated that for domestic purposes "electric current is a far more convenient method than kerosene lamps. . . . Its use in cooking and heating is also possible and would provide better hygienic conditions in the community, lead to the preservation of what meager forests still remain, reduce the number of irksome domestic duties to the house-wife and cut down the volume of imports of kerosene and coal now widely used for domestic purposes" (1936: ix).

Contemporary economists in Palestine and the United States also agreed that it was mainly due to cost that rural areas remained unwired, given that, by and large, coal and kerosene for light and heat were cheaper. The case of Nazareth seems to show how cultural and economic considerations were linked in a way that influenced the priorities of town officials and slowed their readiness to contract for electricity. Nazareth was connected in 1934, yet Faris found that it "would have been served sooner, were it not for financial reasons. The municipality concluded the terms of agreement with the Corporation, whereby the municipality was to extend a loan of £2,000 to the Palestine Electric Corporation toward defraying the expense of building the high-tension line. . . . However the pressing need for water in the thirsty town delayed the project . . . still considered a luxury there" (Faris 1936: 66).

A commercial model determined the forms and directions that the grid assumed and, in such circumstances, connection was largely dependent on the financial considerations of the Electric Company and the financial capabilities of any single local government. On the one hand, the company did not want to invest in high- and low-tension wiring without sufficient guarantees of ample demand. On the other hand, small towns lacked both the capital to invest and the ability of residents to consume electric energy. Kline reports that in Canada solutions to the problem of rural electrification were based on collective contracts with rural communities, aided by heavy government subsidies for the construction of grids. In the United States as well, the extension of the grid to rural areas was mediated through rural cooperatives and "distribution associations" that collectively invested in grid construction (Kline 2000: 133–134, 141–146).

In the United States, the commercial model of grid connection either created or enhanced the gap between rural and urban. In Palestine, it was the difference in the ability of Arabs and Jews to jointly pull resources for investing in the grid that marked them apart during electrification. The Government of Palestine did not provide any subsidies for this purpose. Jewish farmers and Jewish colonies had access to cheap credit and investment capital, but Arab farmers and Arab localities lacked both (Nadan 2006: 58–59). Jewish rural communities—especially those in proximity to already existing high-tension wires—were therefore better able to organize and negotiate electrification on a collective basis. The end result was that, while many Jewish settlements were connected and lighted, many Arab towns and most Arab villages remained in the dark.

The difference between connected and unconnected areas deepened as one moved eastward, further away from the relatively densely populated areas along the Mediterranean coast. The Mavromatis concession, in effect despite earlier British attempts to abolish it, covered Jerusalem and other localities (e.g., Bethlehem) within a radius of 20 kilometers around the Church of the Holy Sepulchre. Ramallah, for example, was therefore connected to this limited grid in 1936. Other major towns in the area, which was years later to become the heartland of Palestine (nowadays familiar as the West Bank) remained unconnected, located far away from the major high-tension lines that fed the urban areas, water-thirsty citrus groves, and nascent industries along the coast. The important town of Nablus, for example, was wired only in 1957, with the establishment of a local municipal electricity project. Before that, its access to electricity, like that of all other towns and villages in the area—to the extent that electricity was available at all—relied on private or collectively owned generators. Looked at retrospectively, then, the electric grid that began its expansion in Jaffa in 1923 effectively drew the boundary that years later would mark the suggested partition line and later the cease-fire line and the (still contested) political border between Israel and Palestine.[61]

And yet this chapter has moved farther away from the assumption that the character and direction of the electric grid were determined by the tensions or the power differentials between a "Jewish-Zionist" electric company and the opposing forces of "Arab nationalism." The chapter began with a description of two models of connection and the creation of two corresponding regions on the same grid. These models later applied throughout Palestine, but it was the model mediated by the local government for joining that grid that served as an almost exclusive method. This way, the grid produced a difference between wired and unwired locations and specifically a difference between wired Jews and unwired Arabs. We are looking at a case of coproduction: assembling a physical electric grid and building a material infrastructure of difference between Arabs and Jews. Jewish-Arab relations indeed tell us something about the process of electrification, but that process tells us even more about the evolution of these relations, stabilizing critical patterns of difference and patterning a trajectory of a growing divide between Arabs and Jews.

3

Wired

EARLIER I DESCRIBED two fundamental differences that were produced by the flow of current on the grid—a regional difference between types of connections (individual and collective) and a difference between connected and unconnected areas. In Palestine, these differences enacted and affirmed a materially visible divide between Arabs and Jews. The expanding grid also created seemingly more subtle differences—between those who were already connected and those who were yet to be connected and between already illuminated and yet-to-be-illuminated streets (Figure 9). An essential property of these differences was that they were destined to disappear with time. They were only a matter of a lag between desire and actual connection. Yet at least for a while, and sometimes for a very long while, this not-yet principle of difference triggered performances of exclusion and inclusion that distinguished streets, neighborhoods, and communities from one another, shaping the meaning of public space and affecting the material and symbolic value of houses and their inhabitants. Proximate access to the grid acted just like other mechanisms of stratification in which social mobility depends on temporality: by no means marginal to the understanding of the construction and consequences of difference!

Consider a handwritten letter of September 6, 1925, concerning the absence of poles and public streetlights, addressed to the Electric Supply

FIGURE 9 *Stringing new wires*
SOURCE: BNA CO 1069/731.

Department of the Township of Tel Aviv and signed by seven residents of a single apartment building. The residents specified the exact location of their building and described their unfortunate predicament: poles and wires were laid out to their right, but ended just before reaching their corner; poles and wires were also laid out to their left, but stopped short just before the corner. The residents were caught in the dark between the lights around them. "We wrote the Electric Company several times," the letter continued, "but they advised us to turn to the [Township] because it was all up to your decision." A township official wrote at the bottom of the letter: "We should issue an order to the Electric Company to position two poles between pole number 532 and pole number 295. A sketch is also needed."[1] A sketch was prepared and approved by the township some time later and sent to the Electric Company for execution. Eventually, the circuit was closed: poles, now numbered 285 and 318, made their appearance on the street.

Thus, the residents of this building joined the grid, leaving behind their former position as yet to be connected for the established position of already connected. With this, they also (re)configured their identities, formally becoming electric consumers under a newly evolving techno-legal electricity regime. In fact, the deeper we probe into this seemingly mundane process, the closer we are to treating it as rather dramatic: connecting to the grid, newly constituted electric consumers (re)joined society.

It is by now commonly recognized that attachment to the Internet and the mobile phone can become as addictive as that to drugs and alcohol and, more generally, that individual subjectivities have become radically dependent on the ability to remain online. But the experience of dislocation and loss that arrives with situations of disconnectedness is not merely psychological. It is sociological through and through, an effect of the distinct character of some networks of circulation, of which the electric grid is an elementary form. Electric grids (and other grids such telegraph lines and wireless communication networks) are significantly different from railroad and highway or airway systems. On these latter connecting platforms, people (or their cargo) are transported from one point to another and they are the circulating particles. On the electric grid, people are not the circulating particles; rather, they function as essential contact points; through their home and office sockets, they are a vital part of the very form and range of the grid, just like streetlights or high-tension wires. Another way of saying this is that the grid uses human contact points just as much as humans use the grid. And it is this aspect of becoming a contact point, not simply a moving particle, that turns the electric grid (or the Internet) into a potent social force. The sense of exclusion that comes with being unconnected, disconnected, or yet to be connected is beyond the reach of the psychological terminology of addiction; it becomes a matter of being outside society and its normative regimes.

Assembling an Electricity Regime

Access to wired electricity was recognized as an essential public good long before the technology arrived in Palestine. The electricity concessions granted to Rutenberg were already treating the electric supply as an object of government, to be governed and regulated by law. Joining society by becoming subjects of an electricity regime may well fit with a formal-legal view. From this perspective, the terms and conditions of attachment to the grid are set in law, and the story of connection is a story of government regulation. At first, electrification was governed by the terms of the concessions and by newly instituted subjects, categories, and distinctions. The Auja Concession for the Jaffa District defined the Electric Company as a "public utility body under government control" and contained clauses whose purpose was to protect "the public" and to ensure the accountability of the concessionaire.[2] It distinguished among four types of electric supply, dividing the universe of the grid into "private light," "public light," "water supply," and "irrigation and industry." It then established the maximum rates that could be charged for each such "domain," relying on the already standardized method for measuring the production and supply of electricity on the basis of kilowatts per hour (kilowatt-hours).[3] The concession also subjected the electric works to the financial auditors of the state and to the technical supervision of the Department of Public Works. And it tied the electric works to other networks of governmental concern, stipulating that the concessionaire should conform to antimalarial regulations and to rules designed to prevent the pollution of rivers.[4]

The concession even contained an antidiscrimination constitutional principle, establishing that the concessionaire should "not show any undue preference to any company or person."[5] It is possible to show, from this formally legal perspective, that it was the law (at first manifested in the concession) that granted the Electric Company legislative powers of its own, establishing that it was "entitled to make rules and regulations to be observed by the consumers of electrical energy."[6]

So the law created the Jaffa Electric Company and assigned it duties and rights, and the law created and regulated the domains of the electric universe, and the law, it seems, constituted the legal category of an electric consumer. From a formally legal point of view, in short, the construction of the material electric grid, the technical aspects of its expansion, and the social effects of connection and distribution all took place within the framework of the law. Such a bird's-eye view suggests an uneven terrain where laws and regulations that lie above, control and determine the ground-level technical, administrative, and practical aspects of electrification.

In contrast, a methodology of a "flattened topography" (Latour 2005: 174) asks us to unfold the terrain, not to privilege state law but rather to look for the ways it is linked to other practices, procedures, routine actions, and participating actors. It asks us to level laws and regulations with technical solutions, municipal interventions, consumer expectations, and material devices; it asks us not to decide—at least not a priori—what matters more and what matters less.

Backed by law, state officials asserted their authority all along the wire: from licensing the production of electricity to planning and construction, from the interaction of poles and wires with their environment to the regulation of rates, safety procedures, technical standards, and operational routines. Their fingerprints may be traced on wiring sketches that they approved, on electric works plans they authorized, and on legal documents they signed. Yet their interventions were often after the fact and more often low-key compared to those of other actors and actions. Many issues had already been regulated elsewhere, assuming shape through the actual experience, work routines, and already established practices of the engineers and technicians who constructed the grid and oversaw its expansion.

This regulation-next-to-law involved a universe of relations that were established "on the grid" during construction, bringing together Electric Company engineers and technicians and a host of municipal officials, engineers, police officers, electricians, and individual consumers. Considered from the point of view of the Mandatory State, electricity was a national concern, a countrywide affair subject to government regulation and control. However, considered from the point of view of the expanding grid and the actual experience of current flowing in the wires, electrification was a messy mixture of local idiosyncrasies and hybrid authorities. A view from above, through the language of law, may focus on how an emergent electricity regime struggled to assert uniform order and regulate relations among actors who were supposedly already there (arranged in given categories and having clear identities). A flat-level view again and again stumbles over instances where the work of assembling a grid created new relations and constantly produced differences and distinctions among new actors.

Local authorities that contracted for light and power played a key role in the assembling of the grid and its electricity regime. Similarly to the established routines that Otter (2008) accounts for in his study of municipal infrastructures in England, the Electric Company and town planning committees cooperated in grid design by coordinating street breaking, positioning poles, allocating land for generator stations, and in general ensuring the technical standards of various construction activities. Local authorities also played a key role in the maintenance of the grid and in facilitating its growth. Municipal employees assumed responsibility for reporting faults and damages to company technicians and engineers; the latter, in turn, were responsible for maintenance and repair. A July 14, 1925, letter from the Electric Company to the Township of Tel Aviv states, "One of your employees climbed a pole and had been seen working between the upper wires. He could have easily got hurt. We definitely forbid your employees to undertake such works. It is our job to perform electric works, at your expense. If such actions would be repeated in the future, we shall be obliged to transfer supervision over the distribution system to our Company at your expense."[7]

The maintenance of streetlights represented a particular concern within the overall emergent regime of wired electricity. The type and quality of the lamps were given serious consideration. Not only residents but also the Department of Public Works and the Township Police were involved in determining candlepower. At times, the township replaced 40-watt lamps with 50-watt lamps in order to brighten a particular area. At times, for reasons of cost, it removed 75-watt lamps and replaced them with 50-watt lamps. The German firm Siemens was also involved, offering different types of streetlights and volunteering advice on types that best fit a town like Tel Aviv "according to our experience with coastal towns."[8] Occasionally, the police asked for increased candlepower in order, for instance, to provide better lights for drivers on busy streets. Occasionally, residents complained of discrimination, arguing that a nearby street enjoyed stronger lighting.[9]

In Tel Aviv, a committee—consisting of the head of the Municipal Department of Light, a municipal electrical engineer, the chief of police, and an Electric Company representative—was in charge of developing working procedures for routine checks of streetlamps.[10] It prepared periodic "light boards" and standard "repair forms."[11] The light board was to be in effect for three-month periods, stating the hours of light for each streetlamp according to its location and size. For considerations of cost, and because they were still considered a luxury, only a handful of lamps were switched on for whole nights, and most were not switched on at all during full moons.[12]

It is remarkable that the Police were part of the emergent electricity regime. Streetlamps were treated as instruments of policing that policemen had to watch for proper functioning. In Zanzibar, writes Winther (2008a, 2008b), electric light became a measure of security against the invasion of evil spirits after dark.[13] In colonial Rhodesia, public lighting served the purpose of policing the nightly movements of black Africans (Chikowero 2007). Schivelbusch (1995), writing about the industrialization of light in eighteenth- and nineteenth-century Europe, showed that at night streetlights were treated as watchmen. Distinguishing between "lighting of festivity" and "lighting of order," Schivelbusch (1995: 137) explains that the latter "was one of the principles of maintaining law and order in late 18th century, pre-revolutionary Paris. By day, 1500 uniformed police were on the streets. By night, 3500 lanterns achieved the same result. The police budget shows how important public lighting was in the security apparatus: Lighting accounted for 15 per cent of the total, making it the largest single item in the police budget apart from the watch (Le Guet)" (1995: 97). In 1920s Tel Aviv, policemen on night watch were supplied with the light board and repair forms. Their task was to mark down failed lamps and other visible damages such as "broken glass" or "broken armature" and to hunt down occasional saboteurs.[14] They handed the forms over to the municipal electric engineer and then to the Electric Company, thereby establishing a firm organizational connection between light and safety.

Local residents in yet-unconnected areas also participated in establishing links between the absence of public light and personal security. Writing letters and petitioning the authorities for electric connections, some invoked the proximity of "Arab orange groves" to residential areas, implying intimidating dangers looming in the dark.[15] One letter stated that it was imperative to light a street leading to a Jewish immigration center in an ethnically mixed neighborhood because "while a resident of Jaffa could still find his way in the darkened street, a new immigrant couldn't possibly do so without proper lighting, especially on rainy days," also associating the absence of streetlights with the heart of darkness.[16]

My purpose is not to suggest that streetlights necessarily acted as repressive devices at the service of ever expanding state surveillance. In this I am in agreement with Otter (2008), who studied the ways municipal infrastructures in the Victorian era facilitated enhanced vision and ways of seeing. Otter suggests that the history of light and vision in the city, including the shaping of the nighttime urban space, was by and large a history of enhanced personal freedoms and emergent liberal subjectivities.[17] However, it is not necessary to fully embrace this positivistic liberalism in order to concur that the spread of the electric grid "sustained nocturnal perception" (2008: 134) well beyond the greater capacities for policing that streetlamps provided. The methodological principle of a flattened topography is uncomfortable with big leaps in whatever direction. Smaller steps are taken, following the traces left behind by the actors and trusting their own associations and connections. Hence, the theoretical and overarching effects of the expanding grid notwithstanding, the most direct action of the outstretching wires was to reshape the spatial contours of the city and to constantly reconfigure its inner divisions.

The expanding grid created new and shifting dividing lines within the city, setting apart those streets and neighborhoods already on the side of light from those still on the side of darkness. Municipalities that contracted for public light played a key role in determining the pace and direction of the grid's expansion. In the collective model of connection, the Electric Company built an agreed on baseline grid, and any further extension had to be separately ordered and paid for by the local authority (Figure 10).

The practice that was adopted in Tel Aviv, later to be followed by other towns, consisted of issuance of "orders of extensions" that increased the reach of the grid within its municipal areas. The Electric Company and the township worked out detailed procedures for executing these orders. The township listed the streets to be covered, sent its surveyors to determine the location of poles on sidewalks and roads, prepared the necessary sketches, and passed the "order" to the company. The company followed the list and attached a serial number to each pole. This allowed the township to specify whether it wished to have a streetlight attached to any given pole and to indicate the requested candlepower (watt) of each lamp.

As a matter of routine, extensions were planned according to the physical length of newly required wires: sometimes the "order" was for only thirty or forty additional meters, and sometimes it was for a few hundred, although rarely did it exceed half a kilometer. It was then left to the Electric Company to determine the type and height of poles that were needed in each designated place. The company and township authorities would also routinely exchange periodic reports and progress logs. These came in the form of tables that listed street names and the length of extended wiring in each as well as maps that graphically represented the crawl of the grid. Electrification was a prolonged event that was carried out in small leaps, and this meant that at any given time some streets already had light while others waited their turn.

Areas that had been equally dark only a year or two before were reconfigured by the advance of the grid, depending on the presence or absence of poles, wires, and streetlamps. And the darkness that always came to unlighted streets at night became intolerable almost immediately. Actual or prospective access to wired electricity turned distance into

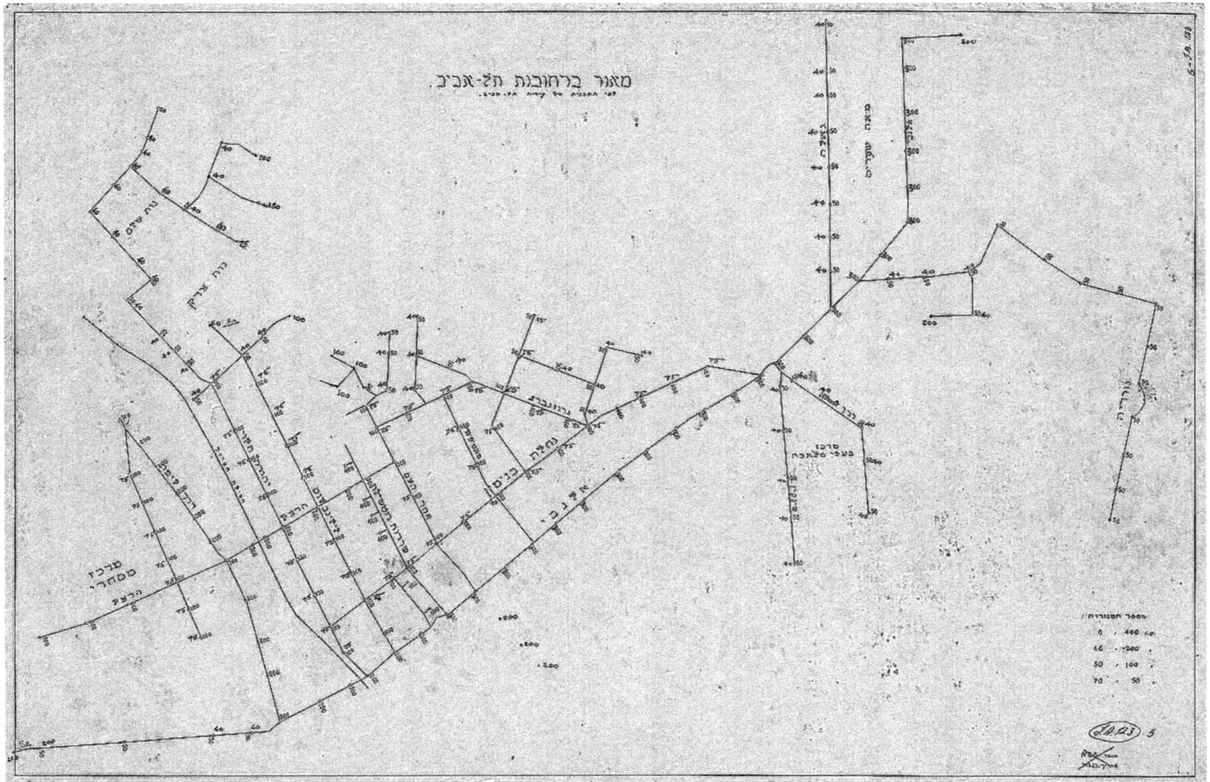

FIGURE 10 *Power line extension into Tel Aviv*

NOTE: This is an early hand-drawn map of the main streets of Tel Aviv, detailing the candlepower of each street lamp (50–100 cp for smaller roads and 400 cp for the town's showcase Rothschild Boulevard). This small grid was the baseline with respect to which the municipality issued extension orders to the electric company in order to expand the coverage of wired electricity.

a relational property: one's belonging (i.e., connectivity) was not determined by geographical proximity to others but by being wired. Then geographical distance lost its social significance as an element on the grid. When not wired, one was "socially distant" regardless of geographical proximity to wires and poles. Perhaps unlike totally unwired communities, the visibly experienced differences between neighboring wired and yet-to-be-wired areas accentuated the tension between expectations and their realization. Disconnection came with an acute sense of exclusion.

Demand soared. An Electric Company report to the Township of Tel Aviv showed that the grid had grown from 12 kilometers of low-tension wires and 2.5 kilometers of high-tension wires at the end of 1923 to over 27 kilometers and 3.6 kilometers by the end of 1925. At that time, the grid also consisted of 12 transformer stations (a total of 1,200 kilowatts). This was a fourfold growth of circulating capacity from the original three transformer stations (a total 150 kilowatts) of late 1923.[18] By early 1925, precisely during this period of rapid expansion, the Palestine Mandatory Government introduced its draft legislation "to facilitate and regulate the supply of electricity for lighting and other purposes."[19] Known as the Electricity Concessions Ordinance, the proposed legislation validated and regulated the division of authority and responsibility among its primary institutional actors with an eye to the protection of public interest. This involved consultations with London: the Colonial Office oversaw the process, the law offices of Burchells provided legal assistance, and the electric engineering consultancy firm of Preece, Cardew & Rider provided technical advice.

However, the insertion of state law and government regulatory authority into the network of electric power did not operate as smoothly and as orderly as planned. It took another two years until, in 1927, the Electricity Ordinance was finally approved.[20] The colonial regulators in Palestine and London soon realized that they could not just enforce a law from above without properly linking it to the local knowledge, working routines, self-regulatory practices, and administrative experience that had been accumulated by the Electric Company. The problem was that the draft ordinance had been prepared along "the lines of the English legislation, and follow[ed] that legislation very closely," adopting the rules and schedules of the English Electric Lighting Act of 1882 and the Electric

Lighting Act of 1899.[21] Pinhas Rutenberg, on reading the 62-page draft, cynically commented that "the number and amounts of penalties mentioned in the ordinance would certainly largely increase the work of the Law courts, but not of the supply of electricity." Legislation in Palestine, he argued, "cannot take the form of an existing law in a cultured and developed country."[22]

Local technical expertise, local awareness of political sensitivities, and local administrative experience—all accumulated during the construction and development of the grid—dictated a different division of technical authority from the one that was envisioned in the draft legislation. Rutenberg opined that some proposed rules and procedures were "technically unacceptable and on the whole impossible to apply with even the best will of those for whom the laws are supposed to be made," that some clauses amounted to "technical nonsense," and that the draft gave the director of the Department Public Works "the right to interfere in things that he will never have the possibility either to control or direct."[23] Rutenberg's reservations received serious consideration by Norman Bentwich, the legal adviser to the Palestine Government. And by the latter's own testimony, they received such consideration precisely "in view of the intimate connection of Mr. Rutenberg with the subject." In fact, Bentwich advised the Colonial Office in London that Rutenberg "did not have the opportunity of studying the draft ordinance itself. . . . It is, I think, expedient to submit his memorandum to the Secretary of State and in the meantime to postpone the publishing of the bill."[24]

Rutenberg was invited to London to take part in a conference on the details of the proposed Electricity Ordinance.[25] The understandings that emerged during those consultations aligned him with the consultancy firm of Preece, Cardew & Rider, which advised the Colonial Office in the matter. Sharing a view that privileged technical considerations over legal and administrative ones, the consultants also expressed the view that it was "best not to impose English practice on local conditions."[26] A new draft circulated between London and Jerusalem in the summer of 1926, this time reflecting stronger links—embodied in the active intervention of Rutenberg in the techno-legislative process—between state regulation, "local conditions," and expert technical opinion.[27] Nevertheless, the High Commissioner for Palestine advised the Colonial Office that "you may consider it desirable to submit the amended draft to Rutenberg before approving it in order to prevent repetition of the events of last year when the ordinance was not proceeded with owing to his criticism."[28]

It would therefore be more accurate to say that the Electricity Ordinance of 1927 joined the grid on adapting to its local conditions instead of being treated as a majestic device that shaped the way the grid worked and expanded. Other elements and movements were no less crucial: the established routines of municipal government; the technical, administrative, and commercial practices of the Electric Company; the labor of inspectors, policemen, electricians, and draftsmen; and the expectations, frustrations, hopes, and pressures of actual and potential consumers. Rules and regulations notwithstanding, soaring demand on the one hand and expansion through municipal extension orders on the other translated into an ever-growing "waiting list" for streets that were yet to be wired. And where there is a waiting list there are pressures and more or less legitimate requests to jump the queue.

Competition among potential subscribers to the grid naturally ensued. Letters of appeal were written, petitions were submitted, and a plethora of asserted connections between electric current and urgency were articulated. Most petitions made their appeals on "public safety" grounds, linking electricity not only to security concerns but also to more general public safety issues, such as the danger of using old flammable materials for lighting or the perils of stumbling into pits on unpaved and unlighted streets—raising images of muddy and dirty holes awaiting the blinded passerby on dark nights. After years of living with oil lamps, the absence of streetlights became a serious hazard and a pressing urgency.[29]

So the incremental expansion of the grid reconfigured the nighttime urban public sphere and determined the ability of this street or that neighborhood to join the forces of progress and modernity. The presence of streetlights and their degree of intensity (candlepower) on any particular street or at any junction accorded those locations importance, attesting to their centrality or significance in the city. And the process of incremental expansion not only produced a difference between those who already enjoyed light on their streets and those who did not; it produced a difference between individuals and their private properties.

This is the direction in which we are now moving: expansion through municipal "extension orders" for streetlights were always an occasion for growth by adding more individual contact points to the grid. Every newly executed order created new possibilities for connections to individual consumers; every new order allowed more wires that could be branched off from the street and extended to residential and commercial premises. In short, appeals for streetlights were in fact entangled with appeals for individual access to electric energy.

A 1924 petition to the township by residents of an upscale neighborhood of Tel Aviv:[30]

> The [Township] had been good to remember its duties to all its streets and neighborhoods and supplied them all with electric light. Yet it forgot to provide our neighborhood with the same service, and in this it discriminated us and set us worse off even from the small alleys of Jaffa, turning us into second class citizens.
>
> Our neighborhood, which due to its wonderful location was expected to undergo the rapid construction of new residential houses, pensions, and hotels, came to a halt because it lacked electric supply. Our neighborhood, whose lovely sight impresses people by day, loses its grace and value at dusk, when the precinct and its fading lamps assume the character of a distant and misty village wrapped in darkness. And how much damage this involves! We cannot accept the fact that our streets and houses, all readily installed for electric current, will remain without electric light. Winter will soon come, bringing storms and waves to our beach-front streets. Traffic at night would become difficult, the streets as they are only equipped with easily shaken fading oil lamps which every passing breeze threatens to smash to pieces. We cannot accept the fact that we pay taxes like everybody else and do not receive the same services like other citizens of Tel Aviv.

The petition was duly signed by dozens of residents. It revealed the acute sense of exclusion that accompanied the state of being not yet wired. Especially humiliating was the fact that less affluent neighborhoods, even in the old alleys of Jaffa, boasted of access or proximate access to the grid. But the petition also captured the grid's dual performance: "We cannot accept the fact that *our streets and houses* [emphasis added]," the residents wrote, "all readily installed for electric current, will remain without electric light."[31]

Material historians and other students of the social and human sciences have paid considerable attention to the effects of streetlights on urban life. Already, as street lighting was taking place in real time, the gradual replacement of oil lamps by gas-lit technology during the nineteenth century, and later the development of arc lights that flooded European and American cities with day-like brightness, was attracting commentary by poets, philosophers, visionaries, and numerous other social observers.[32]

Less attention was being paid to the role of streetlights in facilitating individual connections to the grid. In fact, from the perspective of the Electric Company, streetlights had limited commercial value on their own. Reporting to its board of directors, the company presented tables that distinguished five categories of electric supply, showing comparisons and evaluations. The tables indicated that energy sold for "street lighting" represented roughly 2 percent of the overall supply of electric power.[33] Even when electricity for street lighting was combined with that for water supply (the other component of municipal consumption), the total roughly amounted to only 10 percent of the total supply. Provision of electricity for private lighting was eight times that for streets and other public areas.[34]

The point here is not that streetlights were unimportant but that their importance was largely due to their role in visibly marking those areas of the city that already contributed to that "ambient sub-orange glow cast up into the sky" (Atwood 2000: 395). Lighted streets invited and lured residents to join the grid. Every new streetlight, shifting the boundaries between already-lit and yet-to-be-lit areas, at the same time shifted the boundaries between those residents who were connected and those who were still waiting.

Company reports showed an increase from 700 consumers in October 1923 to 3,141 in September 1925.[35] Basim Faris (1936) calculated an average growth of 15 percent in the number of individual consumers between 1927 and 1932—all in all a 100 percent growth during these five years. His calculations matched the data of the Palestine Electric Company (which took over the Jaffa Electric Company in 1927), the latter showing a 40 percent increase in the number of consumers between 1931 and 1932 alone (see Figure 11).[36]

Year	Units Sold in KWH.	Number of Consumers connected	Gross Revenue in LP.
1929	3,634,838	9,303	90.847
1930	5,500,976	10,620	110.747
1931	8,707,917	12,029	139.673
1932	11,590,350	15,113	145.512
1933	20,136,839	21,934	221.128

FIGURE 11 *Report of directors, Palestine Electric Corporation, for the year ended December 31, 1933*
NOTE: The table indicates the progress made between 1929 and 1933.

The commercial success of the urban grid depended on its ability to extend its public light visibility and simultaneously use this extended visibility to multiply its branching capacity to individual consumers who would serve as its lifelong trusted contact points. The public features of the grid therefore must be understood alongside its role in constructing and shaping various aspects of the "private." And on this count, too, the material construction of the grid was also an assembly of a techno-legal electricity regime.

The Current Meter

Current meters (also known as electric meters) are readers of standardized units of electricity. They measure the traffic of electric energy on the basis of kilowatt-hours. In this standard method of measurement, each kilowatt (1,000 watts) represents the compound of ampere ("amount" of electric energy) and volt (the "pressure" that pushes it). With time units, the standardization of electric energy measurements and the development of current meters was crucial to the selling and buying of electricity on a mass basis. It allowed the seller to take periodic readings and subsequently charge for the "billing cycle," and it provided the buyer of electricity with a way to keep track of consumption.[37]

The maximum rates the Electric Company could charge were a central issue in negotiations with the government over the details of the emergent electricity regime, resulting in various schedules and decreasing scales, future commitments to reduce rates, and a variety of fees for different types of supply and consumption.[38] The company also established regulations that classified current meters (and respective fees and monthly rates) into four categories: two-wire "lighting current meter" of up to 5 amperes, three-wire "power current meter" of up to 10 amperes, three-wire "power current meter" of up to 20 amperes, and three-wire "power current meter" of up to 30 amperes.

The standardization of electric units was the result of linking the electrical sciences with metrology (the science of measurement). Electric units such as volt, ampere, and watt were thus in and of themselves products of aligning delicate actor networks.[39] The current meter in turn provided a material manifestation of this alignment, acting as a boundary object that provided a common language for the technical and regulatory aspects of the evolving electric regime, wedding of metrology, technology, and law, and linking the result to individual consumers. This is precisely the role of boundary objects: to "bridge social worlds" (Fleischman 2006: 78) and to serve as stable connectors between different spheres of activity.[40]

While the most obvious role of current meters was to determine the price to be paid for the consumption of electricity, their physical positioning meant that they acted as boundary objects in yet another sense. Current meters marked the boundary separating the general distribution system of electric current from the wires that branched off to individual premises, domestic or commercial. They drew a concrete techno-legal boundary marking the point where the "public" ended and the "private" began (and vice versa).

Because current meters were precious and sensitive, the Electric Company and later the Electricity Ordinance defined them as company property. Authorized by the concession to do so, the company vigorously regulated around and in relation to current meters and reserved "the right to carry out repairs or other works in connection with them."[41] The newly connected consumer would pay a one-time installation fee for the meter and the company was "entitled to charge a moderate annual rental for [it] supplied and installed."[42] Later, this arrangement changed so that current meters were installed free of charge and a monthly rental charge was incurred according to their type.

Also, because meters were a crucial meeting point as it was, the government had a vital stake in them. While preparing the draft Electricity Ordinance, the High Commissioner for Palestine wrote the Foreign Office in London, asking permission to approach Egyptian electric companies "in order to assist the government of Palestine to assess proposals of the Palestine Electric Company."[43] Upon being granted permission, the Office of the High Commissioner sent five Egyptian electric companies a seven-item questionnaire.[44] Four of the questions concerned electricity rates and application and connection charges. The remaining three focused on the current meter: "What are the rental fees for meters?" "Are charges levied on consumers for special readings of the meter?" "In the event of meters reading inaccurately and a test being carried out, what is the percentage of the inaccuracy allowed before the cost of the test is borne by the Company?"[45] All of the respondents dutifully answered within days, some in French and some in English. The response of the Egyptian Electricity & Ice Company (Suez and Port Tawfic) was typical: it charged a "small fee" for the installation of the current meter, did not charge for monthly or biweekly readings, and bore the cost of the test when the meter was "more than four percent fast."[46]

The Electricity Ordinance that went into effect in 1927 gave the current meter its due place, incorporating into law the relevant articles of the Jaffa and Jordan concessions (Figure 12).[47]

Company to connect with consumers' premises.

16. The Company shall undertake the connection of the said electrical system with the consumers' premises and the Company shall be entitled to charge an annual rental for any electrical meters or other apparatus supplied and installed by the Company for the purpose of measuring the amount of electrical energy consumed by each consumer, all rates and charges for such services to be fixed by the rules and regulations to be made by the Company under Clause 25 hereof.

FIGURE 12 *Excerpt of the electricity ordinance, 1927*

The incorporation of current meters into law formalized the boundary between two distinct spheres of electric activity on the grid. And as they marked the contact area where two parties (the company and the individual consumer) established their relationship, they were anticipated as a potential cause of border disputes. Such disputes stemmed from the fact that current meters were prone to inaccurate measurements that could result in higher (or lower) charges than actual consumption merited. Such cases required effective patrol of the border, the "testing of current meters" to establish the accuracy of the meter "on receipt of a written application from the consumer":

> Doubts as to accuracy will not be considered as exempting from regular payment of accounts. The consumer has the right to demand that tests be repeated by an impartial expert appointed by the High Commissioner for Palestine. The cost of each test including current consumed thereby will be borne by the Electric Works if it indicates that there has been an excess of five percent on meter readings; in other cases the cost will be borne by the consumer. In case the test shows that the consumer has been paying an excess of more than five percent, the excess will be deducted from the last measured account.[48]

The Electric Company also reserved the right to establish the exact location of the current meter on owners' premises, promising to consider owners' wishes but also stipulating that "in the event of a higher cost being entailed thereby the difference in cost will be charged to the consumer." All electric installations from the main distribution system to the current meter were defined as falling within the company's property rights. The company would lay out its wires all the way from the nearest street pole to the current meter on the owner's premises (at the latter's expense). The technical mode of connection between the distribution system and the consumer remained at the discretion of the company. The owner of the unit was responsible for installing at its own expense electric devices and "private" wires from the current meter to the connecting unit.

However, the role of the current meter—beyond its immediate one as a measuring device—had less to do with establishing property rights and more to do with marking a boundary between the public and private spheres of the grid. In fact, the Electric Company made sure that it could easily cross that boundary while still ensuring that meters marked the point where the private domain began and where a difference between the public grid and its individual extensions were constituted. To that end, the company established some rights over the privately owned premises. It made rules concerning "the erection and testing of that part of the installation lying beyond the meter"; it kept the right to "examine connections, installations, current meters, lamps, motors, transformers, wiring, switch apparatus, fuses, and all electrical fittings" of the main distribution system "at all times whether it is its own property or the property of the consumer." It also required the consumer to allow "free access to all rooms wherein electric fittings are installed for the purpose of carrying out such examinations"; to demand repairs at the owner's expense; and to sanction consumers for the use of electrical wires, lamps, and other accessories that could pose a risk to the main distribution system.[49] At the same time, a consumer who transgressed the boundary by bypassing or ignoring the current meter was deemed to have committed a grave offense: "illegal use of current" was defined as electric current taken "without passing through a measuring apparatus," making any such bypass an offense subject to civil and criminal sanctions.[50]

The current meter also participated in shaping the emergent profession of skilled electricians. It marked a boundary—stretching beyond the meter in the direction of the general distribution system—that an "unauthorized" electrician was not allowed to cross. One could and should hire electricians for performing work indoors, but only electricians employed by the Electric Company were licensed to handle the current meter and the wires and connections that led from it to the "public sphere." The company bred and nurtured an elite professional group of its own electricians, one that enjoyed the prestige of working for a hi-tech company that paid relatively high salaries and offered opportunities for overtime pay while other electricians were unemployed. On one side of the current meter were relatively secure yet nonunionized electricians with access to the grid. On the other side were electricians trying to unionize and equally share the work that was available to them on "private" premises. However, the "electrical section" of the General Union of Hebrew Workers was practically useless, leading one union organizer to complain that the employees of the Electric Company looked down on the "section" and rendered it weak by their lack of interest in its actions.[51]

So it was precisely at the location of the current meter that relations were established and identities affirmed. The principled power of the Electric Company to determine the character of these relations was grounded in the original concessions that gave it authority "to make rules and regulations to be observed by consumers of electrical energy."[52] Save for rates, which the government determined in consultation with the company, it was the company that acted as the prime regulator of its relations with consumers. And it was around and with respect to current meters that a substantial part of this form of private regulation was organized.

The current meter designated a newly emergent electric consumer as a subject of rights and obligations. From the point of view of the grid, the consumer—marked as an element of the grid by the meter—was a crucial element in the connectivity necessary for its expansion; not unlike poles and wires, newly constituted electricity consumers joined a material assembly and hence gained the capacity to allow changing amounts of electrical energy to flow through their wires. At the same time, they were something more than material pipelines allowing the passage of electric current through their privately owned wires. Becoming an electric consumer entailed a potentially lifelong connection to a set of regulations that recognized consumers as legal bearers of rights and duties and as agents "free" to control the amount of electric energy circulating on their wires.

The relationship between the Electric Company and the electric consumer cannot be captured in terms of traditional types of exchange between sellers and buyers of commodities. In the case of electric consumers, exchange is mediated through a grid of wires that at once facilitates transactions and depends on them for its very existence. First, this means that the transaction is enduring and constant, thereby simultaneously assembling a sustainable grid and a lifetime consumer. Second, the transaction (i.e., the selling and buying of kilowatt-hours of electricity) did not exhaust itself in any single exchange but was concerned with making the consumer an element of the grid.

The response of the Electric Company to the draft Electricity Ordinance emphasized the crucial need to maintain the simultaneous assembly of the electric grid and electricity consumers. Article 41 of the draft forbade the company from interfering "with the manner in which electricity is used by the consumers." The company objected on grounds that "the consumer may use electrical appliances for which his wiring has not been calculated, thus leading to great disturbances in the whole distribution net." The buying of electric current could not be considered a private matter after which the buyer could use the commodity on the basis of individual discretion alone. Current was constantly circulating, potentially enabling the buyer to breach the boundary, to have an impact on the main distribution system, and therefore to violate the grid's integrity.

The grid created—by means of the current meter—distinct public and private spheres. However, this foundational assembly did not mean that the consumer could be fixed; consumers were neither mere subjects on the private side of the grid nor mere objects (i.e., contact points) of the overall distribution system. Likewise, the grid could not be said to exist independently of its (human) contact points. There were no definite relations of causality and determinacy. On the one hand, the consumer was a product of an "objective" material connection to a grid; on the other hand, this materiality depended on one's "subjectivity": a willing attachment to the grid and a readiness both to consume electricity and to become a bearer of rights and duties. The constitution of the electricity consumer, then, may be better captured by considering the irreducibility of social identities on and in grids: at once elements attached to a network and actors whose connection allows such networks to circulate their "materials."[53]

The Stratifying Grid

The current meter—beyond which lay the web of individual connections—also produced the grid's capacity to maintain asymmetry among consumers. The main distribution system was basically egalitarian. Once its primary arteries (high- and medium-tension wires as well as transformer stations) were in place, it offered a principled equal opportunity for all to consume electric current and to become lifelong electricity consumers.[54] Beyond the current meter lay the private sphere, where asymmetry began, naturalized as a mere function of differential individual consumption. Asymmetry among consumers began as soon as they were connected, depending on the amount of electricity they consumed and on the variety of purposes (and means of displaying those purposes) for which they used it. And with differential readings of consumption came the ability to take part in group formation.

Groups cannot exist by the sheer inertia of a stabilized and endurable "stock of connections" that merely has to be occasionally "represented" by symbolic means. Lamps and electric fans or electric stoves were not, as Latour would have put it, "simply the hapless bearers of symbolic projection" (Latour 2005: 10). They were not (only) a medium of symbolic representation for the display of "progress," "innovation" and "invention," the "urban," the "modern," or "upper-class prestige." Rather, they might be *constitutive* of such categories and attributes.[55] To endure, groups must constantly be performed—to be given opportunities to enact themselves and to be remade anew through attachments and connections that are not in themselves necessarily social.

It was through the measurements of current meters that company officials, government officers, economists, and sociologists collected statistics on electricity consumption. Individually, differential amounts of consumed electric energy "beyond the current meter" allowed for clustering operations: the production of knowledge that was arranged and presented as the expression of group formation. Most concretely, the consumption of electric energy on an individual basis acted as a marker of class position, as a practice that allowed aggregate clusters of consumers—statistically categorized on the basis of their kilowatt-hour consumption—to be enacted as distinct classes.

The first systematic compilation of such statistics in Palestine seems to have been carried out by economist Basim Faris in the 1930s. Faris was primarily concerned with the mostly rural "lower classes" who had yet to be convinced of the advantages of being wired. In the case of these unwired populations, Faris's economic approach assumed that what was at stake was a competition of cost between electric energy and the techno-cultural aspects of kerosene and coal used for light and heat. Yet Faris also analyzed aggregate readings of current meters in the already-wired urban areas and subsequently distinguished among three main groups: the "unskilled laboring classes" who could not afford to wire their homes,[56] the "lower middle-classes" who used only small amounts of electricity for light, and the "upper middle class and wealthy groups" who could consume "freely" and used electricity for purposes other than light.

The Electric Company counted 31,684 wired lamps among private consumers in Jaffa and Tel Aviv in 1925.[57] Faris, roughly ten years later, noted a "small domestic appliance field" (1936: 252) apart from that for light: "the upper middle class and wealthy groups" consumed electric energy for appliances such as washing machines, water heaters, stoves, and electric fans. All of these, because of cost, were beyond the affordable lifestyle and budget of the "skilled workers and white collar people, who fall mostly into the lower middle class."

In the United States, home economists had a role in persuading customers (in rural settings and specifically female homemakers) to consume electric energy beyond "a single drop light" and to use current for refrigerators, electric ranges, heaters, food grinders, and similar appliances (Kline 2000: 150). The use of electricity for purposes other than light, writes Kline, marked a difference between urban and rural populations (2000: 150, 178, 181). In Palestine, the "rural" marked a difference between wired Jewish and unwired Arab localities. However, electric appliances acted as performers of class distinction within cities, distinguishing the "wealthy classes" from the "lower middle class."

The measurements of current meters not only allowed for distinctions between one class and another; they also allowed the creation of knowledge about the social dynamic of grid expansion. Identifying the small scope of the domestic electrical appliance field, Faris was able to conclude that the wealthy classes were not large enough in numbers to sustain the expansion of the grid. In contrast, it was precisely the rapidly increasing demand for residential lights among the "lower middle classes" that resulted in the growing number of branching-off extensions of low-tension wires. "The Jaffa-Tel-Aviv district, and especially Tel-Aviv," wrote Faris, "has a large number of the skilled workers and 'white collar' people, who fall mostly into the lower middle class" (1936: 59). This class, he maintained, used current only for light, yet it was its size that accounted for the more rapid growth of the grid in the Jaffa–Tel Aviv District than in other areas. The ability of the grid to expand, the engine of its movement, depended on the quantity of new contact points and not on the amount of electricity consumed per individual connection. Each new connection allowed the grid to expand; each new connection was a performance of lower-middle-class vitality, in fact affirming and consolidating its very existence as a viable sociological category. Yet again, as we will see in detail in the next chapters, ethno-national distinctions underwrote the class-based observations of Faris. In the Jaffa District, the increase in the number of skilled and white-collar residents was mainly the result of Jewish immigration from Europe.

. . .

The laying out of high- and low-tension wires, the erection of poles, the construction of transformer stations, and the installation of current meters advanced a straightforward commercial logic of expansion—one that sought to transform the consumption of electricity into a universal and taken-for-granted activity equal in its inevitability only to the air we breathe and the water we drink. Still, the grid could not just appear whole and in its full integrity in one instant. It had to win over competing technologies and traditional habits, to discover new outlets, and to constantly increase the number of contact points.

The process of expanding and stabilizing the grid also relied on linking its materiality to a techno-legal electricity regime. Its creators did not simply obey the law but were actively involved in shaping and, moreover, making it. Faithful to a method of inquiry that does not a priori privilege any particular entity or instrument, I showed the grid to be a heterogeneous actor network of technical, administrative, and legal links and associations—a techno-legal arrangement that involved the cooperation and coordination of state law, municipal orders, and technical standards and routines, all embodied in a heterogeneous amalgam of human and nonhuman participants.

The process of wiring, now understood in terms of a techno-legal assembly, highlights the role of the grid in making and marking distinctions, boundaries, subjectivities, and groups. Poles and wires, streetlamps and washing machines, current meters and kilowatt-hours—by now almost transparent to the eye—seem to realize and allow in their materiality some elementary processes of social formation. To illustrate this point, I particularly focused on the current meter as a regulatory instrument. First, I emphasized its role as a node whereby a more or less stable boundary between the public and private spheres of social life could be performed and affirmed. This is not a trivial role once we accept that no such domains as public and private exist save for what is constantly enacted and maintained as such in numerous sites of social action. To build a "general distribution system" and to branch to individual contact points is therefore one constitutive instance of the public and the private.

Second, the current meter scripted principles of difference among the wired. It was through and around it that relations were born and executed, assigning duties, rights, and responsibilities, and assembling a newly constituted "electricity consumer." And it was "beyond the current meter," in the grid's private sphere, that stratifying performances could be carried out by measuring units of kilowatt-hour, compiling aggregated statistics of consumption, and assigning them to different consumer groups. In this way, current meter readings provided data that validated categories of class and allowed distinctions among them. Yet again, this is not a trivial matter once we stick to a leveled terrain and trace how social categories and their differential social attributes are built in action and on the ground.

4

From State to Market

A MAJOR PART OF THE LOGISTICAL SUPPORT available to the British Army making its way from Egypt to Palestine during World War I was provided by hastily constructed railway lines and the use of old Ottoman lines connecting Egypt to Rafah, Gaza, Be'er Sheba, and Lydda, the latter located in the heart of Palestine. By the end of the war, Palestine had become loosely networked by three different types of railway (each having a different gauge): the Sinai desert military lines, the highly strategic Hedjaz line, and the short but vital Jaffa–Jerusalem line (see Figure 13). Their fate had to be decided: they all needed coordination, repairs and upgrades, maintenance facilities, and the settling of legal and administrative issues with respect to ownership and control. The question of electric train traction had also become an issue.

The case of Palestine is telling because the grid had to find outlets on a relatively empty terrain. Electricity had to find a direction, and "direction" is the term that Hughes uses when discussing the goals of electric systems. Direction, he writes, is more important for "a young system than for an old one because momentum provides directed inertia for the latter" (1983: 15). It is one thing to develop a sense of direction when operating in an environment whose material and demographic infrastructure already nurtures an appetite for electric power, and quite another to perform at a place that lacks such an infrastructure. Technological advances

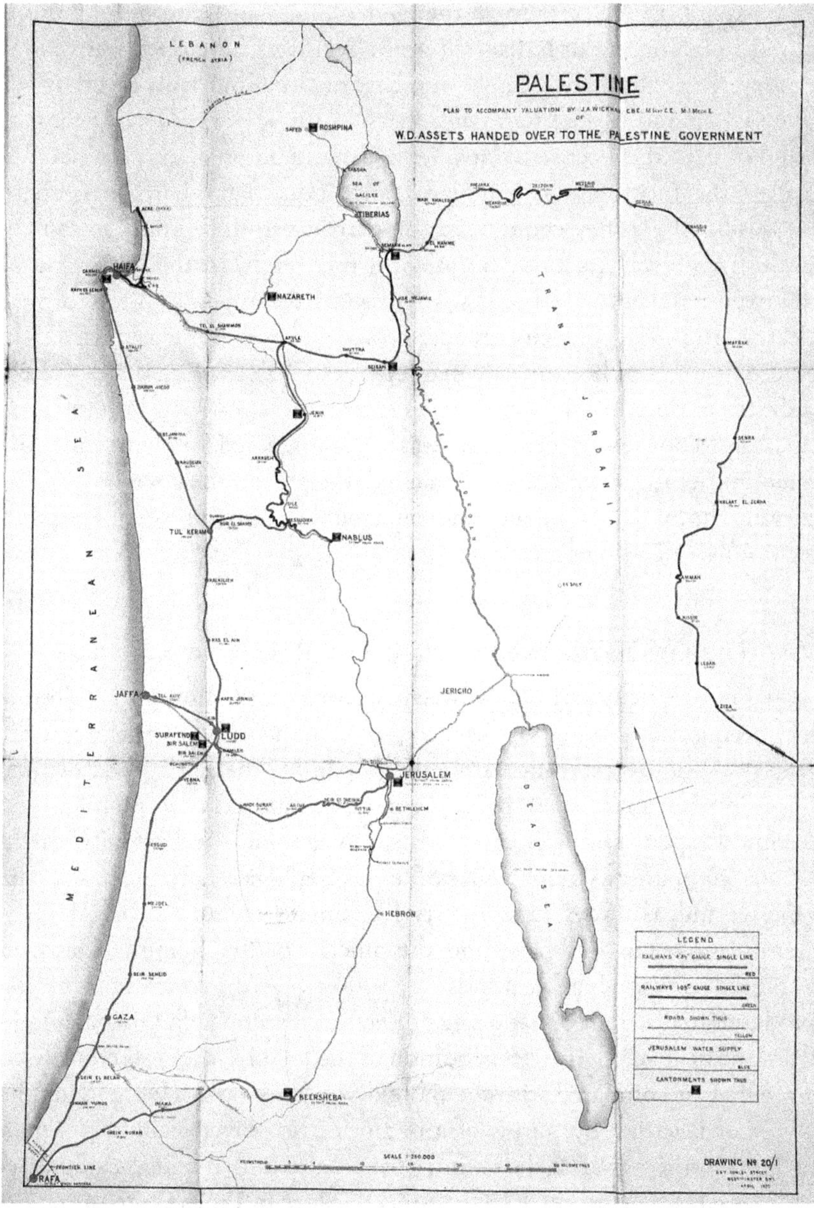

FIGURE 13 *Palestine Railway, 1922*

and inventions in electricity in the United States and Europe all presupposed a high degree of industrialization, relatively developed transportation systems, and considerable numbers of urbanites who could afford to be wired and turned into consumers. In Europe and North America, Hughes says, the electric companies concentrated on supplying electricity to "the most heavily populated and industrialized urban centers" (1983: 15), where they could count on demand from a variety of sources (industry, railway traction, commercial uses, etc.). In the beginning of the twentieth century, observes Kline, electric companies directed their electric grids "to light city streets, power streetcars, and run factories" before turning to the domestic market (2000: 132). Having neither heavy industry nor urban tramway systems to lean on, electric current in Palestine had to flow elsewhere. There was, or so it seemed, only one potential major outlet for wired current: attaching the grid to the government and turning the State into a major and dependable large consumer of electricity through electric trains.

The Grid That Never Was: The Palestine Railways

Projected connections between electricity and railways in Palestine had been in the making for years, ahead of the actual construction of the Jaffa powerhouse and long before the Jordan hydroelectric plant became operational in 1932. These proposed connections had been mobilized by technical experts and scientific studies, topographical and actuarial calculations, comparisons of machineries and technologies, entrepreneurs, state officials, and, as always, legal experts. Exchanges between the British High Commissioner in Jerusalem and the officials of the Colonial Office in London went into high gear in 1921. At stake were the concessions for the production, distribution, and sale of electricity in the Jaffa District and in Palestine in general. The electrification of the Jaffa District was considered a small affair compared with the more demanding task of electrifying the whole of Palestine by means of a hydroelectric power plant that would be built in the north of the country, where the Jordan and Yarmouk rivers met (this is hereafter referred to as the Jordan Project). Yet a no less important issue preoccupied the Colonial Office at the time: the future of railway networks in Palestine and, more broadly, in the Middle East.

Railways were a crucial object of state knowledge all over the British Empire, with experts, models, and strategic plans circulating among almost all of Britain's overseas colonies. Rail networks were counted among the most essential strategic infrastructures both for trade and for military capacities, and ownership or effective control of lines, technology and maintenance, and administrative management occupied British officials at the highest levels.

The Government of Palestine inherited large portions of the highly strategic Hedjaz railway that had belonged to the Ottoman government and had been seized by France and Great Britain during the First World War.[1] At the time, the Hedjaz's main line ran from Damascus in Syria to Ma'an in southern Trans-Jordan. It was designed to reach the holy city of Medina in the Hedjaz (Saudi Arabia), but by the end of the war the section between Ma'an and Medina had become derelict.[2] Another section, operative since 1905, ran between Haifa in Palestine and Daraa in southern Syria, where it joined the main line (this section was known in Palestine as the Yizrael, or Ha'Emek, named after the valley through which it ran).

Because this line ran through the British-controlled territories of Trans-Jordan and Palestine, through the French-controlled territory of Syria, and potentially through parts of Saudi Arabia, the Colonial Office had a substantial stake in establishing the terms of its control.[3] Early discussions during the 1921 Cairo summit resulted in a principled decision to establish a Joint Hedjaz Railway Control Board composed of British and French representatives and representatives of the Hedjaz (Saudi Arabia). However, in practice the French appropriated the parts of the line that ran in Syria, forcing the Government of Palestine to pay for each train that crossed into their mandate territories. It subsequently became clear that the French government in Syria would object to any British attempt to obtain control of the line and to complete it so that it would run all the way from Haifa to Medina under the control of the state-owned Palestine Railways.[4]

Finding it hard to control and develop the Hejaz Railway, the British government in Palestine turned its attention to the old and much shorter Jaffa–Jerusalem line (hereafter JJRW), which ran from east to west right in the middle of the country, connecting the Port of Jaffa to the colonial administrative centers and the holy religious sites in Jerusalem. Roughly midway on the route, near the town of Lydda, the line offered a long-range connection further south, through Rafah and Kantara all the way to Cairo. The Lydda junction therefore became an important strategic site, where the British had built an airport (present-day Ben-Gurion International Airport) and located the British Army and the Air Ministry headquarters (in Sarafand).

The line between Jaffa and Jerusalem had been built by a French company (Société du Chemin de Fer Ottoman de Jaffa à Jérusalem et Prolongements) under a concession from the Turkish Ottoman authorities of Palestine.[5] Steam-powered trains began to operate on it in 1892. In 1914, the line was seized by the Turkish government—at the beginning of the First World War—and remained under its control until November 1917, when Palestine was taken over by the British Army in its advance from Egypt. Under postwar arrangements agreed on in the Treaty of Sèvres, the British had to respect the interests of the line's French owner and operator. At stake were compensations to the French company and the purchase of the line by the British Government of Palestine. Negotiations between a French representative and government officials failed and, according to the terms of the treaty, in such a case arbitration took place to determine the amount to be paid to the French (including calculation of profits of the British Army and later the civil administration from 1917 onward).[6] During arbitration, the Colonial Office authorized a monetary settlement of £600,000, and the line was subsequently purchased from the French company.[7]

The JJRW was maintained and operated by the British military until October 1920 and from then on by the internationally sanctioned British Mandate authorities of Palestine. The High Commissioner treated the line as the core artery of Palestine and made special efforts to ensure that it would remain state property, controlled and managed by the state-run Palestine Railways.

Unlike the Hedjaz line, the future of the JJRW from the start was firmly linked with the prospects of its electrification. The intensive exchanges between High Commissioner Samuel and Pinhas Rutenberg resulted in perfect agreement: both sought to secure a London-approved government commitment to the line's electrification. The messages of the

High Commissioner to the Colonial Office with respect to this issue bore a remarkable similarity to the plans and ideas that Rutenberg had put on the table. Subsequently, the exchanges that circulated between Jerusalem and London in 1921 were unambiguously explicit about the crucial connection between the success of the Jordan hydroelectric project and railway electrification.

Rutenberg, writing to Samuel, declared that electrification of the railways in Palestine was "an essential condition of the realization of my Jordan project"; with such a deal, he wrote, the Electric Company would secure "the creation of a large consumer of energy."[8] A letter from the High Commissioner to the Colonial Office shortly afterward emphasized that "it is an integral part of the scheme that the Railway between Jaffa and Jerusalem should be electrified and that the electric energy for the line should be supplied by the Concessionaire."[9] A follow-up letter from the Government of Palestine's Department of Commerce and Industry to the Colonial Office warned that the Jordan Project "cannot be carried out without the electrification of the Palestine Railways."[10] Another expert report reiterated these opinions, establishing that demand for electric traction was vital "as without revenue from the railway department at the outset of the scheme the financial aspect is insecure."[11]

A crucial principle was established, firmly linking the feasibility of producing large amounts of electricity through hydroelectric energy to the availability of a large consumer (Palestine Railways) that would be able to generate sufficient demand.[12] This was in line with developments elsewhere in Europe and the United States, where the electrification of interurban trains was in full swing, with the expectation that they would be faster, cheaper, and cleaner (Kline 2000: 118). Thus, obeying the imperative of finding a direction for the flow of electric current, the very future of electricity in Palestine hinged on a contract with the government.

More than principled opinions were circulating between Jerusalem and London. Rutenberg carefully collected technical and financial data from the Administration of the Palestine Railways, electric railway journals abroad, the British Institute of Electrical Engineers, and the English Electric Company; from Sir Francis Barker, director of the electric consultancy firm of Messrs. Vickers Ltd.; and from the electric consultancy firm of Messrs. Westinghouse. Moving between Palestine and London,

Rutenberg organized meetings and conferences and eventually secured letters of support from Colonel Holmes, the director of the Palestine Railways, from the director of the Commerce and Industry Department, and from John Snell, the London-based expert who was asked to give his opinion on the plan.[13]

In May 1921, Rutenberg put on the table a detailed plan that began to circulate among the High Commissioner, the managers of the Palestine Railways, officials at the Colonial Office, and the London-based electrical engineering consultancy firm of Preece, Cardew & Rider.[14] All parties wrote memorandums, exchanged opinions, debated the technical and financial data, and engaged in intensive consultations. At stake was a competition of cost between two technologies: the prevailing method of steam (coal) and the emergent technology of electric traction. Rutenberg's document was meticulous in arguing the case for electricity. His report pointed out the latest triumphs of electric traction in Switzerland, Italy, Spain, Norway, France, and even the United States, where in spite of the fact that "coal [was] cheap and abundant" the system was in the process of passing over to electric traction.[15]

A list of technological and financial advantages followed:

1. 20 percent more load-carrying capacity owing to saving of "non-driving weight of locomotives, tenders, water, and coal."
2. Smaller number of locomotives per line owing to the fact that "steam locomotives boilers need more frequent repairing." For example, "On the Chicago-Milwaukee-St. Paul Railway 42 electric locomotives have replaced 112 steam engines."[16]
3. Time of stops at stations was reduced because "there is no necessity for the locomotive to take in water and coal."[17]
4. Considerable savings due to the disadvantages of steam traction: "standby losses in coal" and general "wastefulness in the consumption of fuel," high cost of repair of boilers, expensive lubrication, need for water, and bigger staff.[18]

Rutenberg also cited detailed evaluations of the annual cost of operating steam traction, calculating the price of coal, water, maintenance, and depreciation. He then offered evaluations of future circulations of goods and people on the Jaffa–Jerusalem line and calculated the cost of shifting to electric traction (at a suggested rate of 1 piastre per kilowatt-hour if only line was to be electrified at the rate of 0.6 piastre per kilowatt-hour for the whole of Palestine Railways). All in all, the calculations showed substantial and guaranteed annual government savings with the shift of technology from steam to electric traction even after investing the initial capital needed for the project. The concrete proposal was for the Government of Palestine to finance the electrification of the JJRW for an estimated cost of £250,000 (Rutenberg provided the government with tenders from both AEG and Siemens in Germany) and then commit to purchase electricity at the annual sum of £40,000, to be paid to the Electric Company for its supply of traction power. The Colonial Office, provided with such detailed calculations, was put under considerable pressure to authorize the electrification. The High Commissioner implored the Colonial Office to ensure that the consulting engineers gave serious consideration to the option of electrifying the railways, also making sure that Winston Churchill was part of the loop.[19]

Keeping up the pressure, a follow-up dispatch to London also urged the Colonial Office not to look at the issue only from the "purely railway point of view."[20] London was asked not only to consider the value of competing traction technologies but also to embrace the point of view of the electric grid. From this perspective—above and beyond the interests of the Electric Company to secure future revenues—one had to consider the overall development of Palestine: "By purchasing electric power from the Jordan hydroelectric plant, the Palestine government would ensure the existence of enough money for the Rutenberg plant to electrify the private sectors of Palestine and thereby develop both agriculture and industry" (Reguer 1995: 695). So, even prior to the granting of the concessions, let alone prior to actual construction, some firm attachments seemed to have been making steady progress. The JJRW, on paper, had already been projected as a two-tiered connecting platform: first between the electric grid and the government and, second, between the electric grid and overall economic development. Yet it is precisely the character of some connections that make them behave as mediators whose "input is never a good predictor of their output" (Latour 2005: 39). An eager and capable entrepreneur and a favoring High Commissioner notwithstanding, the participation of experts and different types of data ultimately produced a failed network.

With all the data and arguments provided, the opinion of the consulting engineers, the London-based firm of Preece, Cardew & Rider, was to have a decisive effect. On July 22, 1921, the firm introduced a twelve-page report, "The Electrification of the Palestine Railways" (hereafter the Preece Report) that became the focal point of future deliberations.[21] Expressing a principled positive view of the future technology of electric traction, the report was by far more concerned with costs than with technical issues. It distinguished between "capital cost" (the initial capital needed for electrifying the railways and the trains) and running costs. The capital cost was estimated to be around £250,000, to which had to be added other government expenditures. As for running costs, the report focused on projections of the future price of coal and on the rates that Rutenberg proposed to charge for each "unit" (kilowatt-hour). Financial considerations thereby placed at the center, the report was only mildly favorable of electrification, expressing caution on numbers and projections.[22]

Officials in the Colonial office were also reluctant. First, electricity was not the only technology vying to replace steam. Another competitor, diesel traction, lurked from behind. At least one official at the Colonial Office expressed the view that with the oil soon to be flowing from Mesopotamia, a better and cheaper technology would be available.[23] Second, it was doubtful that the Treasury would approve the initial investment capital: "I am afraid that we should have great difficulty in persuading the Treasury that we were justified in these circumstances in spending a quarter of a million out of loan funds on the electrification of the railway."[24] Third, there was some principled resentment of the fact that Rutenberg was pushing the government into long-range expenditures and commitments. This was not simply a matter of restraining a too eager entrepreneur. Technical deliberations and financial calculations were linked to a broader colonial view of the appropriate relationship between the State and a local commercial enterprise. The British government was favorably disposed to grant concessions to Rutenberg, but this did not mean that electrification could depend on the government binding itself to become a major electric consumer.[25] On the projected grid, the colonial government and a native market were to be kept apart and unequal. Reguer (1995) cites an official who wrote that the "time has come to intimate pretty plainly to Mr. Rutenberg that it is for the Government, and not him, to prepare schemes for the redevelopment of Palestine, and that he really must not ask contractors to submit estimates for works to be carried out by and at the expense of the Palestine Government."[26]

Rutenberg responded to the Preece Report and the reservations it created in the Colonial Office with a new ten-page document. He disputed some of the calculations made by Preece and introduced new, favorable cost calculations based on the number of locomotives that would be needed to run the line several times a day. He also presented a new method for calculating running charges that would reduce the overall annual government expenditure by a further 20 percent.[27] All this to no avail; reluctance to invest the capital needed for the project, compounded with uncertainty and doubts about the annual expenditures that would be involved in running the line, was sufficient to shelve the plan. The possibility that the Electrical Company would purchase the line and invest its own resources in electrifying the system was also rejected. The JJRW was considered a state asset, and the government insisted on its ownership of it. Thus, the Jordan Concession was given a green light to proceed in September 1921, but without an agreement on the JJRW's electrification.[28]

The consulting engineers sent another report to the Colonial Office in December 1921, addressing Rutenberg's response to their earlier report of July. Taking account of the capital cost, the price of electricity compared to the price of coal, and traffic conditions on the Jaffa–Jerusalem line, they concluded that it was not "possible for us to state definitely that if the Railway Administration purchased electricity . . . it would be cheaper than the existing system using steam locomotives."[29] A follow-up opinion from the locomotive superintendent in Palestine, delivered to the general manager of the Palestine Railways, seems to have put the final seal on the attempt to link the railways to the electric grid. The superintendent wrote that the price per kilowatt-hour that was being asked was "still fairly high" and that similar undertakings in India in 1916–1917 supplied power at lower rates. Expressing full agreement with the report of Preece, Cardew & Rider, he added that the expectation of significant traffic on the Jaffa-Jerusalem line was far from certain and that it was not "an ideal section from an economic standpoint."[30]

And then there was silence. Just as happened four years later, when ideas about an electric tramway in Jaffa and Tel Aviv were circulating with no apparent results, all the plans, calculations, opinions, and appeals concerning the JJRW came to an abrupt stop. However, there was a crucial difference between the failure to agree on the electrification of the JJRW in 1921 and the abortive attempt to build a tramway in Jaffa in 1925. In both cases, the prospective investors (commercial entrepreneurs in one case and the British government in the other) balked at the capital costs of electrification and at the future rates of electricity. Yet, while in 1921 Rutenberg was investing considerable energies in securing a deal, expressing a firm belief that the electrification of Palestine depended on it, the Electric Company was quite indifferent, in 1925, even to the modest prospects of electrifying an urban tramway. Something had happened: a shift in the "point of view of the grid" had occurred within a span of four years.

The Government as Electric Consumer: A Tale of Prudence

The first part of this book mainly followed the electric wires from the newly built powerhouse in Jaffa toward the streets, shops, and houses in the Jaffa District. With construction and the beginning of production, the biggest promise and the most visible sign of the grid's vitality was the provision of light and water supply (by means of electric engines) to the towns and bigger agricultural settlements (mostly Jewish) in the region. By replacing kerosene lamps on the streets and persuading a rapidly increasing number of residents to become electric consumers, the grid made steady progress.

However, the demand for public and private light could not sustain massive long-term growth. Other outlets had to be found that would allow the grid to provide electric power on a larger scale. The electric consumption capacities of local municipalities was limited, electrically illuminated streets with high-voltage lamps were still considered a luxury, and most public lights were turned on sparingly. Residential consumption was also very limited, principally confined to low-voltage lamps and a very low number of domestic appliances. While the early failure to secure a contract to electrify the JJRW had not prevented the relative prosperity of the grid in the Jaffa District, future plans to expand still seemed to depend on attaching the government to it. Absent a major contract, the search had to turn elsewhere. At the most basic level, this involved prompting government offices, departments, and other state-owned or state-run facilities to establish attachments. Negotiations to that effect typically involved the government's Department of Public Works, whose position in such matters was often decisive. Some progress was made in this direction, and by the end of 1923 the offices of the Palestine Post, the Palestine Railways, the Customs House at Jaffa Port, and the Government Hospital in Jaffa were connected.[31] The Electric Company also hoped that it would be granted a government contract to supply electricity to the large workshops that were being constructed by the Department of Railways in the port town of Haifa.

The Department of Railways planned to construct its own small electric plant for that purpose, but Rutenberg thought he had identified an opportunity to secure a lucrative contract with the government that would allow him to begin extending the grid beyond the range of the Auja Concession. Backed by High Commissioner Samuel, he approached the chief engineer of the Crown Agents for the Colonies and proposed to supply electricity to the railway workshops with a new power plant that he would build in Haifa almost exclusively for that purpose. He argued that his electric supply, based on German machinery and technology, would cut the government's planned expenditure on the railway workshops by half.

Some intensive negotiations and highly detailed calculations followed throughout late 1921 and early 1922. The general manager of the Palestine Department of Railways, The chief engineer of the Crown Agents for the Colonies, and Preece, Cardew & Rider held somewhat differing views. The technical issues involved various estimates of demand and prospective supply, considerations of converting electric power to direct current, and the future prospects of electric supply by the planned hydroelectric plant on the Jordan River. The general manager of the Railway Department thought that self-reliance would be cheaper, more dependable, and more readily available. The consulting experts thought otherwise: "A public supply company should be in a position to supply electricity at a lower price than it should cost to produce in local workshops."[32] The chief engineer of the Crown Agents for the Colonies, responsible for submitting a conclusive opinion to the Secretary of State, doubted the ability of the Electric Company (which at this stage had not begun operations anywhere in the country) to meet the workshops' urgent demand. Ultimately, the railway workshops were advised to produce their own electricity. Nevertheless, the company soon began building a powerhouse for Haifa, and it could still hope to benefit from the supply of electricity to government facilities in that rapidly developing strategic and industrial town. But the failure to secure the contract with the railway workshops indicated yet again that the government was reluctant to rely on a "native" private company when it came to strategically sensitive installations.

This reluctance to depend on a native grid came into fuller view in 1923. The Air Ministry High Command in Sarafand—about 14 kilometers southeast of the Jaffa Powerhouse—declined an offer to contract the Electric Company and resolved to build its own electricity-generating station. Rutenberg lobbied hard to change this decision throughout 1923, recruiting the High Commissioner to his side. His efforts bore fruit in December 1923, when the Air Ministry agreed to enter into negotiations with his company, marking a major achievement for the company and its eager grid.[33] However, strings were attached. From a technical standpoint, this endeavor involved laying out a high-tension transmission line that would carry a 15,000-volt 3-phase alternating current from the Jaffa powerhouse to Sarafand's Army cantonment. From a financial standpoint, the project required an estimated investment of £12,000, expecting a yearly return of £5,000. The company treated the agreement with the Army as a strategic asset. It committed itself to this costly project and invested considerable effort in finalizing the contract and building the line as fast as it could.[34] Work on the ground notwithstanding, the conclusion of the contract involved lengthy legal and technical deliberations concerning wiring standards and the quality of connections between the generating station and the "transformers, switchboards and all the apparatus necessary for delivering the current to the cables of the Air Ministry in the camp."[35]

Underlying the technical deliberations over the character of the grid was a principled issue of jurisdiction: the type of relationship that the grid would have established with the military forces of the Sovereign. The law had to be brought in yet again, requiring the participation of lawyers alongside technical experts and consultants. Previous chapters documented how the expanding grid drew and regulated the boundaries between public and the private spheres and how it established a techno-legal electricity regime. These techno-legal boundaries aimed to stabilize and formalize the relationship between a commercial company and ordinary civilians, constituting them as "electric consumers." The privileged rights of access to private premises that the Electric Company enjoyed with respect to the treatment of electrical machinery "beyond the current meters," however legitimate in "civil society," was cause for concern when applied to relations with strategically sensitive state apparatuses. The language was often technical, but the intervention of high-level legal experts in London could not be mistaken. During negotiations, their major concern was that the company had too much of a free hand to approach, maintain, or otherwise handle military machinery and instruments on the army's own premises, suggesting that "the point of connection [between the general distribution system and the army's electric equipment] was not sufficiently defined."[36]

A new and special type of electricity regime had to be scripted, one that attached the grid to British laws at home, requiring the Electric Company to conform to the "Wiring Rules of the Institution of Electrical Engineers of Great Britain." British legal experts had other concerns as well, all underpinned by the idea that licensing a non-British entity to electrify a colony was one thing and letting it meddle with the standards and prerogatives of the Sovereign was another. The lawyers and advisers in London raised concerns about the jurisdiction of local courts to decide potential disputes between the company and the Army; they worried about the legal complications that might ensue from the fact the company was not legally registered in the United Kingdom and pondered what the implications would be if the company liquidated.

All of these legal concerns had one thing in common: the presumed relative symmetry between the Electric Company and its consumers could not be applied to its standing vis-à-vis the State. A symmetry had to be maintained, and the grid had to allow for new techno-legal arrangements that would reflect its subjection to the Crown.[37] Yet again electric wires functioned as intermediaries rather than mere mediators, providing a slightly different actor network: a social formation that was assembled from slightly different elements and that resulted in a special electricity regime with its own boundaries, its own rules for crossing those boundaries, and different roles assigned to some of its boundary objects.

One of the most significant benefits of the contract with the British in Sarafand—beyond the obvious one of securing steady income from a large consumer—was that it allowed the Electric Company to extend the grid beyond the original geographical limits that had been projected by the Auja Concession. Norman Bentwich, chief legal adviser to the Government of Palestine, was prudent enough to let London know that, the contract with the Air Ministry notwithstanding, the Auja Concession prioritized the demands of local authorities in Jaffa and Tel Aviv for water supply and street and resident lighting, and stipulated that surplus supply could be used for industry and for the irrigation of orange groves in their immediate vicinity.[38] The concession was silent about extensions such as the one planned for Sarafand. Bentwich therefore suggested that the concession be amended so as to cover supply to military installations. While such an amendment was never formally introduced, facts were established on the ground. Not for the last time, the material grid expanded faster than the law that was supposed to regulate its progress. Formal plans notwithstanding, the grid had a power of its own.

Rutenberg was quick to realize the lucrative potential of the special extension to Sarafand. The high-tension line that exceeded the limits of the original concession ran along some major towns and agricultural settlements, offering the grid new opportunities for extended connections. Specifically, it ran in proximity to the Arab towns of Ramleh and Lydda (as well as dozens of smaller villages) and the Jewish settlements of Rishon Le-Zion, Nes-Ziona, and Rehovot. While the Arab municipalities were slow to act, the Company immediately entered into negotiations with Rishon Le-Zion and soon after with the other Jewish settlements.[39]

Still, the modest contracting with government installations, the ever-growing number of domestic consumers, and the growth in demand from municipalities could not in themselves account for the grid's substantial growth. In spite of its failure to electrify the JJRW—considered in 1921 to be an essential condition for grid expansion—and in spite of its failure to conclude an early deal with the railway workshops, the Electric Company built a new powerhouse in Haifa. Its revenues surpassed original estimates. In 1925—its second year of operation—the Company reported that "the yearly income of the Jaffa Power House has almost reached the estimate for the fifth year of work. The estimate was made in 1922."[40] Revenues flowed in, yet their source was not the one that had originally been projected. The grid had changed direction. Without the government as a major consumer, it had discovered "the market," if it had not invented it outright, and was suddenly able to attach itself to engines, machines, and water pumps in the domains of industry and agriculture.

Lost Links

One trace of the aborted plan to electrify the railways was left behind in a letter sent from the Palestine Department of Commerce and Industry to the Colonial Office in 1921. The letter indicated the vital connection between the viability of the hydroelectric Jordan Project and the electrification of the JJRW, explaining that this was so "owing to the fact that the demand for energy from other consumers is, *and will probably long remain,* only a small fraction of the total supply which will have been made available" (emphasis added).[41] The connection of electricity to trains was vital, then, because Palestine was bereft of other major sources of demand for electric power such as industry.

A second trace was located at the other end of the time tunnel. In 1926, the Electric Company sought to secure a government loan guarantee for the Jordan Project. The request was made under the British Trade Facilities Act, and the immediate authority to which it was directed was the Trade Facilities Advisory Committee. The purpose of the Trade Facilities Act was to encourage investments in and purchases from depressed British industries in England so as to relieve unemployment and enhance production. The request to secure a government-guaranteed loan for the

Jordan Project under the act was flatly refused initially on grounds that "the general prosperity of the electrical trade in England" prevented its application to the case of the Palestine Electric Company. The advisory committee, which operated under the Act, went beyond the formal argument and expressed its views on the commercial feasibility of the Jordan Project. It referred to a 1923 expert opinion that, like the opinion of others at the time, established a firm link between the potential success of the Jordan Project and the electrification of the JJRW. Observing that the electrification of the line had not been realized, the committee doubted the feasibility of the hydroelectric project. It also commented on the Electric Company's own predictions of demand for electricity in Palestine submitted to the government in 1923, observing that the Jordan scheme had predicted that only 26 percent of total electrical production would be consumed by industrial demand, but now, only three years later, it had unexpectedly changed its estimates to 61 percent of anticipated industrial demand.[42] Doubting the new estimates, the advisory committee refused to authorize the government-guaranteed loan on these grounds as well.

Rutenberg—fighting to secure what seemed the most vital interests of the prospective all-Palestine grid that would ensue from the Jordan Project—did not give up. He wrote a long memorandum recalling the British government's discontent earlier in the decade on learning that German equipment had been used for the electrification of Jaffa. He reasoned that a similar situation applied to the Jordan Project: German equipment was significantly cheaper than British equipment, and only a government-guaranteed loan under the Trade Facilities Act would legally bind him to buy electrical equipment in the United Kingdom and tie the hands of his board of directors, which otherwise would opt for cheaper German tenders.[43] This reasoning found favorable ears at the Colonial Office and persuaded the Secretary of State for the Colonies to plead the case again with the Trade Facilities Advisory Committee.[44]

Sir Alfred Mond—an influential Jewish British industrialist and a strong supporter of electrification in Palestine—also pleaded the company's case. He wrote a long memorandum suggesting that Palestine had changed—the Palestine of today was fundamentally different from the Palestine of only two years ago. Expert opinions that were three years old were seriously outdated. Look at the numbers, he wrote: demand for electricity was soaring; heavy industry was established and was demanding large amounts of electrical energy; no less than 160 factories depended on electrical supply in Tel Aviv alone. Palestine was now developing just like the Western United States, with nothing similarly comparable in Europe. The electrification of the railways, he added, had become "unimportant."[45] Another source—referred to by all as "the demand of industry"—was available. Although some officials in the Colonial Office were not as enthusiastic as Mond about "the future of Palestine as an industrial country," they still thought the advisory committee had not attached "sufficient importance to changing conditions" and had given "undue weight" to earlier reports.[46] The pressure proved effective. On September 23, 1926, the advisory committee reversed its earlier decision and approved a government-guaranteed loan of £250,000 to the Palestine Electric Company.[47]

A new link was established, connecting the grid to new sources of demand. The electrification of the railways—perceived in 1921 as an essential condition for the commercial feasibility of the Jordan Project—had become "unimportant" by 1926. It had become unimportant because from 1925 onward the grid was attaching itself to an inflow of private capital, to a flow of commercially oriented immigrants who invested in industry and in new agricultural technologies. And it was this new attachment that had become powerful enough to substitute for the earlier expectation that only an attachment to government resources would allow the grid's serious expansion.

The Flow of Private Capital

An unanticipated wave of Jewish immigrants began to arrive at Palestine in 1924. Roughly 60,000 landed between 1924 and 1926. Relative to the size of the Jewish minority in Palestine (approximately 15 percent of the total population), this amounted to a staggering ratio of 285 newcomers to 1,000 residents in 1925 alone (Lissak 1989: 214).[48] The new immigrants were steered to Palestine by hyperinflation, economic meltdown, and follow-up drastic (and largely ineffective) corrective economic measures in Poland. In January 1924, the Polish Parliament under Prime

Minister (and Minister of the Treasury) Wladyslaw Grabski abolished its collapsed currency and replaced it with a new one. It also launched a major tax reform that included new, heavily imposed property taxes (the Act on State Treasury Repair and Currency Reform). These measures hit the commercial classes particularly hard and drove many to emigrate.

The crisis in Poland coincided with newly introduced restrictions on immigration to the United States that seriously disrupted the decades-long flow from eastern and southern Europe to North America. In 1924, the U.S. Congress passed the Johnson-Lodge Immigration Act, which set new quotas and restrictive caps on immigration and seriously and deliberately curtailed mass Polish immigration to the United States. Driven out of Poland and unable to enter the United States, many Polish Jews turned to Palestine. First, Palestine already had a modest Jewish community ready and willing to absorb newcomers. Second, one could expect to find in Palestine some unexplored economic opportunities now that it was ruled by Great Britain. Third, and perhaps most important from a practical point of view, British immigration policies allowed into the country those who could qualify as "owners of capital" (i.e., who could declare an imported capital of £500 or more). It is estimated that 40 percent of the immigrants to Palestine in 1924–1926 were allowed in under this designation.

The flow of people with modest private capital and commercial and entrepreneurial dispositions mainly aimed for the country's nascent cities. Of the newcomers, 62 percent settled in Tel Aviv, which by the end of 1925 numbered 40,000 residents compared to 12,862 in August 1922.[49] Having been a small Jewish township next to the city of Jaffa only a year or two earlier, Tel Aviv was now showing signs of becoming Jaffa's serious competitor both in size and in the demographic profile of its population.

A surge in the supply (and demand) of public and private light and a corresponding rapid expansion of the grid was the most immediate and visible effect of the flow of people to Tel Aviv. The Jaffa Electric Company's directors' report for 1924–1925 (its second year of operation) showed that the number of connections to the Jaffa grid had more than doubled within a year. At this early stage of expansion (September 1924 to September 1925), the company also reported an increase in "consumers of light" (from 1,528 to 3,141), an increase in residential lamps (from

14,905 to 31,684), and 1,787 new electric meters installed on consumers' premises.[50] A confidential company report prepared toward the end of 1925 showed that between October 1923 and July 1925 demand for "lighting for residences" had grown by 260 percent, "lighting for streets" had grown by 160 percent, and demand for "water supply power" had grown by 290 percent.[51]

The Jaffa–Tel Aviv grid rapidly expanded, boosted by a sixfold growth in electric production capacity.[52] The growth of the grid could be detected in the frenzy of requests for "grid extensions" and in the periodic exchanges between the Electric Company and the Tel Aviv Municipality.[53] At the beginning of 1926, the company issued a full report on the length of the grid in Jaffa and Tel Aviv,[54] which showed that the low-tension grid had 12 kilometers of wire by the end of 1923 and over 27 kilometers by the end of 1925. It also showed that by the end of 1923 the grid was supported by three transformers—each with the capacity of 50 kilovolt-amperes (for a total 150)—and, by the end of 1925, by eight 100-kilovolt-ampere and four 50-kilovot-ampere transformers (for a total 1,000 kilovolt-amperes).[55] Notably, the report also stated that only one transformer was shared by both Jaffa and Tel Aviv, indicating that the overwhelming bulk of the growth was taking place in Tel Aviv and not in Jaffa.

Yet the demand for public and private light, and even the enhanced demand for "water supply power," still consisted of a very small, if not marginal, part of grid expansion. The real growth lay elsewhere. Contemporaries and latter-day observers alike perceived the new wave of immigration in 1924–1926 as one that brought capitalism to Palestine. The combined flow of people, capital (however modest by European standards), and commercial orientation yielded immediate results on the ground. Both investment capital and government revenues doubled between 1923 and 1926. Looking back, one observer declared 1925 to be "the year of industry" (Beilin 1987: 48). Industrial output grew in areas such as food, print, timber, machinery, chemicals, and building materials. The promise of a growing textile industry seemed to have been particularly appealing, given that many textile manufacturers were moving their workshops from Poland to Palestine (Shenkar 1963: 163).

The influx of commercially oriented immigrants and private capital was not limited to urban areas. The mid-1920s were also the time when citrus farming became a booming agricultural industry. In a land where Arab-owned orchards covered considerable tracts of land, the Jewish colonization of Palestine was coupled with the citrus industry from the start. The first grower's guide for citrus farming in Hebrew had been published in the late nineteenth century, offering, on its cover, "theory to practice in planting orchards in the Holy Land" (Fellman 1891). Having learned cultivation from Arab farmers, Fellman advised readers to master Arabic and its particular citrus-related jargon. By 1891, orchards surrounded Jaffa in all directions. All Jewish agricultural settlements in the vicinity, with Petah-Tikva first and foremost, cultivated citrus groves. In the mid-1920s, Jewish immigrants to Palestine began to vigorously plant new orchards all along the coastal plane north of Jaffa, almost all the way towards Haifa 100 kilometers distant. Thus, alongside the flow of immigrants to urban areas, mainly to Tel Aviv, the 1924–1926 wave of Polish immigration resulted in the establishment of new agricultural settlements where Jewish farmer-entrepreneurs sought to modernize the technologies of citrus cultivation. The number of Jews living in agricultural settlements grew by 75 percent between 1924 and 1929, and agricultural output more than doubled in the process (Ziman 1929). Most of these settlements had been established by farmers with a modernist outlook on agriculture as "a distinctive capitalist area of private initiative" (Karlinski 2001: 328).

Private capital—invested as it was in industrial and agricultural enterprises—depended on access to electricity. The expansion of the electric grid, on its part, became increasingly dependent on industrial and agricultural demand. Securing these networks of interdependence would substitute for the earlier reliance on government contracts in general and the electrification of the railways in particular. Rather than the engines of locomotives, the grid could now expect attachments to industrial engines and machinery and to electric pumps needed for irrigation.

The transition from state to market was not without its difficulties. On one layer, the attachments of the grid to industry and agriculture involved some bitter disputes with both the colonial government and Zionist institutions. On another layer, electricity had to fight its way against competing technologies that were cheaper, more accessible, or simply already established. These two layers of interface with the electric grid were related to each other because technological issues were often at once tied to questions of regulation, duty taxes, credit, legality, scientific and administrative surveys, and organizational capacities. The actor network of the grid had been redrawn, establishing new links with actors such as agricultural surveys, pumps, industrial engines, and orange trees.

5

Assembling a Jewish Economy

THE LETTER FROM THE JERUSALEM OFFICE to the London headquarters of the Zionist Executive was exceptionally angry. It concerned the position of the Jaffa Electric Company on a proposed amendment to the government's Customs Exemption Ordinance. The amendment exempted from import duties buyers of small "prime movers" (electric or steam engines) who intended to use them for generating power in their factories without having to rely on a wired connection to the electric grid ("the general distribution system"). The letter opened with a recap of a meeting between Pinhas Rutenberg and Frederick Kisch, a senior official of the Zionist Executive: "I enclose a note on a somewhat astonishing interview which I had yesterday morning with Rutenberg, who came in and out of my office with that astounding combination of resemblance to a steam-roller and a whirlwind which he alone is capable of achieving," he wrote, and "his intervention in the matter of the Customs Ordinance is 'thoroughly unjustifiable' and 'intolerable.'" Kisch explained that many "small industrialists" were in dire need of that exemption and that the Zionist Executive was lobbying the government to introduce it. Yet Rutenberg, Kisch complained, "is ignoring the interests of the country" and is "thinking only of his own project."[1] These allegations were not unfounded.

The Electric Company had approached the High Commissioner and demanded the cancellation of the exemption. Rutenberg explained his position to the Zionist Executive:

> [You insist that] I should withdraw my demand for cancelling the exemption from custom duties of power producing machinery. . . . I strongly object to the fact that the Palestine Executive of the Zionist Organization was conducting negotiations with the government [on this issue] without previously consulting us. . . . Whilst I cannot have any objection to have the advice of such a competent and authoritative in this matter person as Sir Alfred Mond, I regret not to be able to agree to Sir Alfred Mond being the arbitrator in this question between the Palestine Zionist Executive and our company, this involving an important issue of principle that our company should in its relations with the government in connection with their interests have a third party. I cannot create such a precedent without previously consulting my board of directors.[2]

A month later, aware that the intervention of the Electric Company had created a storm among the members of the Zionist Executive in London, Rutenberg wrote another letter, this time using a more conciliatory tone: "I was convinced that this clause was introduced by some of our enemies in order to fight the interests of the Concession. Had I known that the Zionist Executive was engaged in negotiations about such an important question, I would have discussed with the Executive this matter before taking any steps with the Government."[3]

So the Zionist Executive, operating as a proxy for "small industrialists" and promoting the goal of Jewish industrial development in Palestine, strongly embraced the amendment to which the Electric Company objected. The latter considered the matter as no less than a war waged against it. It viewed such an exemption as a measure designed to discourage factory owners from becoming wired electric consumers and encourage them to prefer what would be a cheaper alternative. Duty-free "prime movers," in short, were regarded as enemies of the grid. At least for a while, it was the position of the Electric Company in this matter that was adopted by the government. In a June 30, 1924, meeting, the government discussed the Customs Exemption Ordinance:

> A communication had been received from Mr. Rutenberg requesting the Ordinance be postponed, pending a decision from the Colonial Office in respect of a clause exempting from customs dues prime movers, which he wished to have omitted. Mr. Rutenberg had communicated direct with the Colonial Office. The council decides to publish the Ordinance in the next Gazette omitting the clause in question in accordance with the instructions of the Secretary of State, when the ordinance is promulgated.[4]

It would take more than a year before the government finally resolved to recommend that the Secretary of State exempt from import duties "prime movers, machinery and the component parts thereof, imported for use in factories and industrial establishments by or for the proprietor or manager thereof."[5]

The Electric Company acted in a way that—from the point of view of the Zionist Executive—ran against the interests of the Jewish community in Palestine. From an institutional Zionist perspective, it was necessary to use all measures possible to facilitate Jewish industry by reducing the costs of production and easing access to engines and machinery. The matter looked different from the point of view of the expanding grid. To generate demand for wired electricity, it was necessary to successfully compete with fuel-based machinery, which could slow the grid's expansion and reduce company revenues. And if anticompetitive measures were needed, so be it.

Only three years earlier, a report by the government's Department of Commerce and Industry confidently predicted that demand for energy from private, commercial, and industrial consumers was "*and will probably long remain*, only a small fraction of the total supply which will have been made available [emphasis added]."[6] A small fraction it may have been, but the grid nonetheless had to secure whatever advantages it could harness to its aid so as to attach industrial demand to wired electricity. Objecting to the exemption from custom duties on imported prime movers was a principled measure that coincided with novel developments on the ground.

. . .

The previous chapter taught us that contact points between the general distribution system of the grid and its consumers were marked by current meters. And we have seen that the measurements of current meters in standardized units of kilowatt-hour allowed for comparing levels of residential consumption and for the creation of aggregated clusters of different "classes."[7] Yet the method of kilowatt-hour measurements for establishing not only individual but aggregated degrees of electricity consumption also allowed for other types of comparable measurements (see Figure 14). A slight detour is in order here to elaborate a bit on the nature of the grid as a producer of social formations.

		1932	1933
		Kwh	Kwh
1.	Private Lighting	1,945,000	2,727,000
2.	Industrial Power	4,060,000	6,576,000
3.	Irrigation	4,400,000	9,029,000
4.	Water Supply	910,000	1,464,000
5.	Street Lighting	275,000	340,000
		11,590,000	20,136,000

FIGURE 14 *Report and balance sheet, Palestine Electric Corporation, for the year ended December 31, 1933*

NOTE: The table is a detailed statement of the electrical energy sold in 1933 as compared with that sold in the previous year. It classifies the sale of electric energy into five categories, the largest of which is irrigation. Chapter 5 follows this classification to its sources and offers links between electric energy and other economic developments.

Comparing levels of consumption and assigning them to different categories or classes is not a trivial matter. Such categories are a product of a more or less stable network: "a series of elements with well defined relations between them" (Mol and Law 1994: 649). There must be numbers at work (kilowatt-hours), and there must be reliable instruments that produce these numbers (current meters) and people who periodically attend to these instruments and know how to read them and record and compile the readings, and people who know how to fix or recalibrate them when they fail. And then there must be some offices that receive these scattered readings, where there are people and more or less definite procedures for compiling, processing, and computing all these measurements, eventually presenting them in the form of different and comparable categories or classes of consumption. This is an insight with which drafters of actor network theory equip themselves in setting out on a study: the universe in which comparable categories (here of consumption and consumers) are formulated and then realized through stable measurements does not exist in the order of things but is rather an effect of the network. The type and number of categories according to which data are collected and processed may change over time. Data may be arranged and rearranged, for instance in order to improve their informational quality or to provide a more accurate picture of the reality of electrical consumption out there. And when the network works without systemic failures, using the flow of current to produce various comparable categories, then we may say that the grid becomes a maker of social formations.

The Auja Concession prefigured three categories of consumption: "street lighting and water supply," "electric lighting other than street lighting," and "irrigation and industrial purposes." These corresponded to the expected trajectory of the grid even before actual measurements. Based on experience elsewhere and on projections of limited supply, the main purpose of these categories was to establish different monetary rates for each consumption type. The first category lumped together street lighting and water supply, tacitly assuming that the consumer in such cases would be a municipal authority. The second category was phrased "negatively," perhaps testimony to the limited value that was projected for domestic consumption. The third category lumped together industrial and agricultural demand.

An Electric Company confidential report in mid-1925 announced a "phenomenal growth of activities."[8] It also arranged the newly compiled readings of kilowatt-hours according to four categories of consumption: "lighting of residences," "lighting of streets," "industrial power," and "water supply power." This slightly new categorization no longer lumped together municipal consumption, probably because the figures were already large enough to justify a distinction between public lights and water supply. The category of "irrigation" disappeared altogether, leaving the demand for "industrial power" to stand on its own. The numbers spoke for themselves: while demand for the first three categories increased by an average of 200 percent, demand for industrial power grew by 800 percent over a time span of 21 months.[9] With it came an emphasis on a broader generic distinction, splitting the universe of electric supply into the hemispheres of "light" and "power."

The Electric Company released figures that indicated a dramatic proportional change between "light" and "power": "The proportion of energy sold in January 1924 has been for light 48 percent and for power 52 percent (industry and water supply) whilst in April it was for light 26 percent and for power 74 percent (industry and water supply)."[10] The appetite for "industrial power" kept rising steadily. The company reported on industrial demand for 68,464 kilowatt-hours in July 1925. This demand further rose to 106,374 kilowatt-hours (another 50 percent increase) only two months later and to 202,155 kilowatt-hours in early 1926 (almost another 100 percent increase compared to three or four months earlier).

The network of measuring kilowatt-hours established and clarified the rising importance of "industrial power" as a viable social category. "Industrial power" became a power to be reckoned with, which was especially surprising because no one had expected it only a couple of years before. In 1926, the Jaffa Electric Company became part of the countrywide Palestine Electric Company, and from then on company reports provided data for the supply and consumption of electrical energy on a nationwide basis.[11] By 1926, the company had sold over 2.3 million "units" (kilowatt-hours)—of which 60 percent went to industry—showing steady progress and dramatic leaps in supply and demand over the next nine years (Figure 15).[12]

Year	No. of Consumers Connected	Units Generated (in kWh.)	Units used at Corporation's Works (in kWh.)	Units Sold (in kWh.)	Gross Revenue in L.P.
1926	6,550	3,048,442	182,851	2,343,764	66,791
1927	7,477	3,143,442	183,045	2,527,126	71,315
1928	8,582	3,890,622	413,664	2,973,701	79,900
1929	9,303	4,930,960	747,113	3,634,838	90,847
1930	10,620	7,199,317	852,162	6,168,198	125,582
1931	12,029	11,530.198	845,432	8,707,917	139,673
1932	15,113	16,629,054	499,714	11,590,350	145,512
1933	21,934	29,120,155	756,719	20,136,839	221,128
1934	35,397	47,485,776	888,072	34,385,515	363.900
1935	53,246	71,885,097	3,151,352	50,362,193	488,443

FIGURE 15 *Report and balance sheet, Palestine Electric Corporation, for the year ended December 31, 1935*

The categories, the measurements, and officially published company data told a territorial story that referred to Palestine as a single political and administrative unit. There was nothing in the numbers that revealed anything more about consumers apart from their classification as "big" and "small" or as residential, municipal, or industrial. Other traces only appeared sporadically and mainly in non-numerical forms. Company reports linked the growth to Jewish immigration: "The country is enjoying unprecedented prosperity due to the continued influx of Jewish capital accompanied by Jewish initiative and enterprise." Yet "speaking for the grid," the company retained a territorial view, preferring to downplay the significance of the emerging difference between Arabs and Jews: "The prosperity is not confined to the Jewish settlements, [but is rather] shared by the entire population."[13]

Economist Basim Faris, comparing the electric markets of Palestine and Syria in the early 1930s, noted the "rapid growth of the industrial market for electricity in the territory of the Palestine Electric Corporation" (1937: 153). He relied on company data that showed that electricity for industry was more than double the supply for light on the territorial basis of Mandatory Palestine. Faris also tacitly recognized that Jews used electric power for industry much more than Arabs did, yet he nonetheless insisted that "there seems to be no explanation [for the phenomenal growth in demand] other than the favorable rates which compete with substitute sources of power" (1937: 153). It seems that Faris missed, or did not wish to emphasize, the attachment of the grid to the flow of fresh Jewish capital from abroad.

There is no doubt that it was statistically possible to arrange the data so as to match the categories of consumption with Arab and Jewish sources. Yet the incentive to do so was low. Instead, the company aligned itself with the statistical territorial fix of the Mandatory State. The government balked at statistics that would be arranged on an ethno-national basis. In 1928, the government conducted its "First Census of Industries for 1927," which included "all factories and workshops, regardless of size, that produced for the market" and followed the territorial principle in analyzing and representing the Palestinian economy in an aggregate nationwide form. Only many years later did it emerge that the survey originally distinguished between Arabs and Jews and contained data that validated Palestine's divided economy. Faced with these politically alarming statistics, the government shelved the publication in an effort "not to further aggravate the already tense Arab-Jewish relations" (Metzer 1998: 155; Gross 1979). The first census of industries was eventually released in

1929 without the statistical analyses that attested to the widening economic gaps between Jews and Arabs (Elisar 1979, cited in Metzer 2003).

Years later economic historian Jacob Metzer was still able to reconstruct and confirm the links between electric power for industry and Jewish demand. He compared the "ratio of electric power (in kilowatt-hours) used to output produced" in Jewish industry to the territorial figures. For the former, he relied on a publication of the Jewish Agency. For the latter, he relied on the government's "statistical handbook" and its "survey of Palestine." The comparison showed that the overall Palestine ratio of electric power used for output produced in the manufacturing sector grew fourfold during the 1930s while the ratio in the Jewish sector grew substantially faster (Metzer 1998: 157–158).[14]

Such traces and reconstructions indicate that the category of electricity for "industrial power" in effect embodied the attachment of the grid to "Jewish industrial power." Electric current mainly flowed to Jewish industry, following it to its nestled locations in Tel Aviv and Haifa and more generally to the newly arrived Jewish capital that accounted for this dramatic rise. The number of small-scale factories and industrial workshops in Palestine more than doubled between November 1923 and July 1926. Tel Aviv experienced the fastest growth. The town hosted 93 factories in 1924 and 262 in 1925; of the 42 textile factories operating in the country in 1925, 30 were also located in Tel Aviv. By comparison, the port town of Haifa had only 52 factories in 1925. While Tel Aviv attracted the most people and small workshops, Haifa seemed to have attracted a bigger share of heavier industries and corresponding capital investments.[15]

In 1924, the Electric Company began construction of a new diesel-fueled powerhouse in Haifa. It acquired the land it needed for the project, imported the machinery, and completed the construction and the laying out of the distribution system at the extraordinary speed of less than a year.[16] By the end of 1925, and with even greater intensity in 1926, the electric grid covered most of the Haifa bay area, also connecting Acre. Just as in the District of Jaffa, supply for street lighting and domestic purposes had become marginal relative to power for industry. In fact, industry demand became so intense that machinery had to be moved in order to meet it: the company announced that "in view of the unexpected increase of demand for energy" in the District of Jaffa, a British-made

Fraser & Chambers electric engine had to be "loaned temporarily from the recently opened Haifa Powerhouse."[17]

In Haifa, just as in Tel Aviv, the grid catered its major bulk of electric current to the new industries. Pressing the British administration to authorize a loan for the Jordan Hydroelectric Works, Sir Alfred Mond explained that electric supply in Haifa was delivered by diesel engines of about 4,000 horsepower and this was insufficient because "there is a cement works demanding 2000HP and salt works demanding 400HP" alone.[18] Analyzing 1920s conditions in Haifa, sociologist Deborah Bernstein finds that industrialization was largely Jewish–capital intensive and that the local Arab population "could not equal" the know-how, capital, and institutional support that was enjoyed by Jewish industries (2000: 117). All in all, although Tel Aviv had by far more workshops and factories, the demand for electric power for industry was roughly the same for both cities by the early 1930s.

As the grid expanded in the direction of Jewish industrial capital, it established a material techno-statistical platform for the assembly of what gradually became known as the "Jewish economy." This assembly, in turn, required more than electric power for industry. It required sustained and systematic statistical efforts to collect, measure, compare, and report on market activity by compiling figures on an ethno-national basis. This was not a trivial endeavor. To become a reality, the "Jewish economy" had to be disembedded from within the broader national economy. Its assembly had to be fixed and stabilized against other efforts of fixation and had to engage in a politics of calculation that would set it apart from both the colonial government and the economic activities of the Arabs.

Whether national or ethno-national, an "economy" does not exist in the order of things but is an effect of certain frames and measurements. An economy does not exist independently of the sciences that define and measure it, but is rather embedded in these sciences. Economics, writes Michel Callon, is a discipline that is at once "a stakeholder" in certain activities of exchange and a contributor "to their construction and their shaping" (2005: 16).[19] Timothy Mitchell suggests that the "economy" was also embedded "in social theory, in colonial administration, in national income accounting, and in all the other practical activities of mid-twentieth-century social life that formatted, shaped, and made the new field of national politics" (2002: 118). An economy, any economy, writes Mitchell, requires a "politics of calculation" (2002: 8).

Mitchell attributes perception of the economy as a distinct domain of reality accessible to scientific observation to the studies of Keynes in colonial India. After Keynes, he writes, the economy "came to refer to a self-contained structure or mechanism whose internal parts are imagined to move in a dynamic and regular interaction, separate from the irregular interaction of the mechanism as a whole with what could now be called its exterior" (2002: 82). Against the sociological imagination of Karl Polanyi, Mitchell suggests that the economy came into being not by disembedding market relations from a larger social setting that previously contained them but by "embedding certain twentieth-century practices of calculation, description, and enumeration in new forms of intellectual, calculating, regulatory, and governmental practice" (2002: 118).

In Palestine and elsewhere, government departments invested considerable effort in producing data and in collecting annual statistics on issues such as "salaries and prices, monetary transactions and banking, foreign trade and land transfers, agricultural and industrial output, construction, transportation, communication, and local and national expenditures" (Metzer 2003: 10). Such measuring and analytical activities were not only part of the normal functions of the state but also a means of affirming the ontological reality of a national economy, asserting authority over a definite territory, gaining legitimacy, and deploying reasoned and scientifically based public policy.[20] But another type of calculative politics developed in Palestine against the effort to articulate a national economy. The rapidly increasing gap between Jewish and Arab consumption of "industrial power" enabled a material basis for comparisons between the two populations. More specifically, the formally universal grid became a building block in the assembly of a separate "Jewish economy." But a material infrastructure is only a necessary, not a sufficient, condition for such an assembly. To exist as a viable category and a taken-for-granted fact of reality, the "Jewish economy" had to be performed and represented in figures, charts, and statistical data.

This work was assumed by economists and statisticians who enjoyed the institutional support of Zionist organizations. An early bird was the Palestine Economic Society (PES), which from 1921 onward published bulletins, articles, proceedings, compilations, news, tables, and a wealth of statistical data concerning the economic activities of the Jewish population. The Jewish Agency also began to publish industrial and agricultural surveys and analyses in 1923, and another Zionist institution, Keren Hayesod, began to publish the *Statistical Abstract of Palestine*. By 1925, with the creation of the Jewish Industrial Association, the institutionalized production of academic knowledge and public policy papers focusing on the commercial and monetary activities of the Jewish community was already assuming, and further substantiating, a reality of a separate Jewish economy (Metzer 2003). In 1929, the economist Yehoushoua Ziman published *The Economy of Eretz-Yisrael in Numbers*. "Eretz-Yisrael" was the Hebrew name for Palestine, and the book surveyed, measured, and reported on the Jewish economy almost exclusively, already treating its distinct existence as an established fact.

State knowledge that would validate the existence of an ascending Jewish economy could prove that the British fulfilled their commitment to facilitate a Jewish National Home in Palestine by economic rather than political means. Yet such acknowledgment threatened to alarm the Arab majority and prove its claims that the Zionist/British colonization of Palestine did not work for the benefit of the country as a whole. The result was that the Mandatory State tried to preserve its own hold over the economy and to avoid the publication of data that would contribute to economic divisions in the country. The territorial basis of the economy was thus officially preserved, but the result was that, while systematic knowledge about the "Jewish economy" was assembled throughout the mandatory period, only few and scattered data were produced about Arab economic activities. The little that was known was the product of a Jewish economy asserting itself in comparison to an inferior Arab one.

The "Arab economy" concentrated on agricultural products. Its manufacturing output in 1927 was dominated by four single branches: flour milling, tobacco products, soap, and olive and sesame oil. Jewish manufacturing was by far more diversified, relying on larger plants (a 2:1 ratio in favor of Jewish manufacturing in terms of employment per factory), a larger stock of capital (a 4:1 ratio in favor of Jewish manufacturing), and "a continuous increase in its utilization of motor power" (Metzer 1998: 155–156). Following the large inflow of capital from the mid-1920s onward, and given the availability of wired electricity, manufacturing became the largest industry in "the Jewish economy" while agriculture became "the largest single Arab industry." Although Jews were still a minority (less than 20 percent of the population), their economy "generated about half of the country's manufacturing value added by the early 1920s" (its share rising to 80 percent in the 1940s) (Metzer 1998: 154).

In sum, the assembly of the "Arab economy" in Palestine mostly took place by default and perhaps may be better understood as a "negative assembly." Arab institutions, writes Metzer, "did not create statistical mechanisms for the systematic collection and analysis of data." The overall effect was a great "imbalance between the wealth and scope of the economic data that was created in the Mandate period by the Jewish community and about it and the poverty of statistical data about the Arab economy" (2003: 14). Except for an edited book on the "economic organization of Palestine" in the late 1930s (Himadeh 1938), Arab economists and statisticians did not try to assemble an "Arab economy." By and large the "Arab economy" remained part of overall government data except for a few special studies and surveys, some of which were carried out by Jewish institutions and researchers. The "Arab economy," then, mainly existed as a kind of an ephemeral shadow, appearing as the ambiguously inferior "other" of its Jewish counterpart. The Jewish politics of calculation, then, not only disembedded itself from within the "national economy" but also provided a basis for an ethno-national politics of partition.

Power for Agriculture

And then there was water and electric power for water pumps and irrigation and the involvement of the grid in hydro-politics.[21] The Auja Concession did not make much of agricultural water use. It treated electric supply for the "irrigation of orange groves and gardens" as a "surplus" that perhaps would have existed after supplying electricity for other municipal and domestic purposes.[22] "Irrigation" did not merit its own category in the measurement of kilowatt-hours in 1925. By 1932, it had surpassed "industrial power" and had become the largest single category among the five categories that the company by now specified. By 1933, the amount of kilowatt-hours sold for irrigation roughly equaled the amount sold for the other four categories combined (see Figure 14).[23] Just like the category of "industrial power," the "irrigation" category mainly embodied the connection of Jewish agriculture to wired electricity. In turn, this connection became an important material element in establishing Jewish agriculture as the second half of the increasingly "separate" Jewish economy.

The connection between electricity and water for irrigation was solidified with the development of electric pumps. In the United States, water became an object of government early on, writes Patrick Carroll, but before the technological innovations in large-scale irrigation in the early twentieth century it was of concern mainly in terms of water rights (2012: 318). In Palestine, the question of water rights and the technological capacity to tap into a large source of groundwater by electric means came together in the mid-1920s, in what became a battleground of hydropolitics. The case pitted the Mandatory State and the Electric Company against each other, binding and mobilizing into action legal advisers, hydrological experts, colonial officers, and international law.

The Auja Concession, anticipating the construction of a hydroelectric powerhouse on the Auja (Yarkon) River, granted the Electric Company the rights to use the waters for both energy generation and irrigation. Yet the plan to build such a powerhouse did not materialize and electricity for the Jaffa District was produced by means of diesel engines at a different location. Instead of releasing the company from its commitment altogether, the British authorities approved the change of plan on a temporary basis, each time extending the date by which the hydroelectric powerhouse had to be completed.[24]

In 1925, the government announced a plan to exploit the waters of the Auja to supply water to the city of Jerusalem. The Electric Company objected on grounds that it still retained water rights over the river. The dispute soon reached crisis proportions because it pitted two different concessions against each other: the British concession that allowed the company to use the waters of the Auja and the Ottoman-granted concession that allowed Mavromatis to produce electricity and to supply water to the District of Jerusalem. The validity of the Mavromatis Concession and its principled status relative to the Auja Concession had just been decided in the Permanent Court of International Justice in The Hague after a prolonged dispute between England and Greece.[25] A Palestine Electric Company report informed its directors that "in March 1925 the International Court of Justice at the Hague decided that Mr. Mavromatis' Jerusalem Electricity Concession granted to him by the Ottoman government in 1914 should be confirmed and adapted to the present economic conditions of Palestine."[26] The Palestine government, acting on this ruling, granted Mavromatis an authorization to use 11,000 cubic meters of water daily, to be pumped from the Auja for the Jerusalem water supply system (later 20,000 cubic meters). The company intended to fight this decision. Rutenberg claimed that the Auja Concession gave him exclusive rights to the waters of the Auja both for the production of electricity and for irrigating the district and that no other person or entity could exploit them. The government tried to pressure Rutenberg by pointing out his breach of the original commitment to build a hydroelectric powerhouse. In May 1925, Rutenberg again asked the government to extend the permission that he was periodically receiving to postpone the hydroelectric scheme. This time, the government informed the company that it would grant the requested extension only in return for "submitting a scheme to supply electricity to Petah-Tikva" and, in case this supply came from the diesel powerhouse in Jaffa, only in return for surrendering "all irrigation rights in the river Auja."[27]

By this time, in mid-1925, it had become clear that there was no real need to build a hydroelectric powerhouse on the Auja because the electricity that it could generate would be marginal relative to the enhanced capacities of the already operative diesel-fueled station.[28] Having been refused an extension, Rutenberg threatened to sue the government in court. Alarmed, High Commissioner Samuel notified the Colonial Office that Rutenberg "threatens to construct this unnecessary station if this is the only means whereby he can preserve intact his irrigation rights."[29] Water rights, water supply, and prospects of irrigation became objects of dispute that yet again coupled legal and technical issues on the Jerusalem–London circuit. Rutenberg solicited the expert opinion of a water engineer showing that the company would suffer significant losses if waters were to be pumped for Jerusalem: "The Growth of the population in the Jaffa region will necessitate in the near future the withdrawal of large amounts of water from the Auja for the drinking and domestic purposes of this population and any further withdrawal for outlying districts would considerably reduce the water available for irrigation, which is of vital importance for the Jaffa region." The company then raised the issue of lost revenues. Based on its water engineer's calculation that the withdrawal of water for Jerusalem would amount to total of 5,000,000 cubic meters "during the eight months of the dry season," it suggested that this amount was enough for irrigating "at least 10,000 dunam of orange groves" that would yield an income of "about 5000 Pounds a year or a capitalized value of about 100,000 Pounds."

The dispute with the government, then, was not only about the welfare of the Jaffa population but also about the financial welfare of the company. Still, the company made it clear that it raised the financial issue only because a private "third party" (i.e., Mavromatis) was about to profit from the proposed supply for Jerusalem: "We beg to state that whilst our Company would have been prepared to yield to the Municipality of Jerusalem the amount of water required for the needs of its population without any compensation whatever, we do not consider it possible to prejudice the interests of the Company by making such a sacrifice for the private benefit of third parties."[30] Thus, also networked into the dispute around the Auja waters was Rutenberg's concern that allowing water to be supplied to Jerusalem would increase the value of the Mavromatis Concession—a grave concern if the concession was ever to be bought by the Palestine Electric Company.

The government equipped itself with its own expert report, prepared by the director of Public Works. In contrast to the company's techno-financial findings, this report found that the water in the Auja could comfortably suffice for both irrigation and water supply to Jerusalem and that Rutenberg's figures and calculations were either deliberately false or simply wrong.[31] Still, the director admitted that he had no adequate, reliable maps at his disposal and that his calculations were performed under time constraints. Topographical and hydrological knowledge was still scarce, allowing disputing parties to put up figures and arguments that were hard to verify. This may be the reason that officials in the Colonial Office seemed reluctant to let the matter be decided on technical grounds alone and asked their legal advisers to consider whether any other defense might be open to the government.

The Colonial Office asked the legal advisers of the Royal Courts for opinions on two issues: first, whether the Electric Company was in breach of the concession and hence had lost its water rights; second, whether earlier correspondence between the company and the government in which extensions for the construction of the hydroelectric powerhouse had been granted could be construed as an implicit agreement that a final waiver of the obligation to build the powerhouse would be granted in exchange for the company's waiver of its water rights. This line of action brought no relief. The legal advisers explained that the government could still require the Electric Company to fulfill its obligation under the concession (and to build an unnecessary powerhouse) but that "we do not consider that the company has committed a breach of its obligation to erect the power station" because it had duly received government extensions to that effect. The lawyers also informed the Colonial Office that it was quite impossible to construe the correspondence (between the government and the company) as constituting an agreement binding the Electric Company to permit the withdrawal of 20,000 cubic meters of water a day from the Auja.[32]

The British lawyers found eventual refuge in returning to the technical aspects of the plan, opining that the government would not be in breach of the Auja Concession if the permission granted to Mavromatis left sufficient water for irrigating the Jaffa District: "If Mr. Rutenberg elects to test this matter in the courts it must eventually be decided upon technical evidence as to the amount of water available and the amount reasonably required for the purposes of the concession."[33] It was on the basis of this legal opinion—coupled with the report of the director of the Department of Public Works refuting the calculations of the company's water engineer—that a decision was reached. The government relieved Rutenberg of his original obligation to build a hydroelectric powerhouse on the Auja without rescinding his water rights.[34] In turn, it allowed Mavromatis to use the Auja River to supply water to Jerusalem.[35]

The Electric Company defended its water rights in the Auja with a zeal not matched by actions on the ground. Apart from a number of

Arab villages that pumped water for irrigating their orange groves, the relatively large Jewish colony of Petah-Tikva had the biggest potential to become a consumer of large-scale irrigation. By 1925, the company had purchased the land and the infrastructure required for electric-powered pumping stations. In July 1925, it reassured the government that "an irrigation scheme to Petah-Tikva is under way."[36] The company reported that it had completed the "erection of the High-Tension overhead line to Petah-Tikva and low tension distribution system in that colony" in December 1925. It also announced that "new works in connection with the irrigation of the Petah-Tikva region are expected to be commenced during next summer [1926]."[37] It was only in late 1929 that the company submitted an irrigation scheme for Petah-Tikva, but it was yet to be approved by the government in 1930.[38] This finding begs for explanation, especially in light of the fact that the Electric Company had otherwise shown remarkable agility in executing projects in which it had a stake.

Water as a Boundary Object

In 1930, a British Commission of Inquiry discovered the basic reason behind the Palestine Electric Company's slowness in utilizing the Auja for irrigation. The findings of the commission, published and known as the Hope-Simpson Report, established that the profits that the company generated from feeding electrically powered pumps attached to wells trumped its willingness to develop a large-scale irrigation scheme: "A further drawback to the exercise of the Auja concession by the concessionaires is found in the fact that if its water were to be used generally for irrigation of the plantations of the Maritime Plain, the existing demand for electric power to pump irrigation water from wells in the plantations would be liable to diminution or possibly to cessation." The commission could only lament that the greater mass of the Auja water flowed into the sea because of the company's untapped exclusive rights: "It is regrettable that one of the chief sources of irrigation in the country, the Auja River, has been included in a concession" and that "the Government has in one case parted with the irrigation rights in an important source to a concessionaire, and steps should be taken to ensure that in that case satisfactory arrangements are made for a supply of water for irrigation at an early date" (1930: 83, 149).[39]

The mandate and ambition of the Hope-Simpson Commission was to tie the issues of immigration, land settlement, and agricultural development in a way that would allow the government to deploy policies that could serve the country as a whole. The Hope-Simpson Report placed the need to develop a national water regime at the center of the system it wished to create. Water was to play the primary role of a boundary object: "adaptable to different viewpoints, and yet robust enough to maintain [its] identity across different groups" (Carroll 2012: 313, quoting Star and Griesemer 1989)—an object of government around which the interface of political, scientific, administrative, legal, and technological issues could be organized into a unified regime. In California, the role of water as boundary object was important in the emergence of the modern "technoscientific" state (Carroll 2012). In Palestine, this role had a unique part in determining the immigration policies of the colonial government: the regulation of Jewish immigration was explicitly tied to the availability of irrigable land as a measure of the absorptive capacity of the country (Reichman et al. 1997; Alatout 2009).

Sensitive to this link, scientifically based hydro-politics became a site of Zionist activity well ahead of government activity. Jewish hydrologists began to study and survey underground water and the sources and characteristics of surface water. Gross (1987) identified three Jewish hydrologists who immigrated to Palestine from Germany in the mid-1920s. Prompted into action by the dramatic development of citrus farming on the coastal plain, but also looking elsewhere in the country, they began to publish scientific reports and to submit detailed memoranda to Zionist organizations on the water economy of Palestine and issues such as underground water, the size and nature of irrigable areas, and the depth of wells and their level of salinity. Alatout (2009: 371) finds that the link between water and the country's absorptive capacity was thus firmly solidified in the 1930s.

The Mandatory State lagged behind. Analyzing the nonexistent water regime of Palestine, the Hope-Simpson Report noted state failure in both science and law. The government had neither agricultural development policies in place nor sufficient topographic and hydrologic knowledge. Water rights were of an "undefined nature," and the use of water was highly localized. Water was by and large held in private hands, with speculators holding water rights that were sold "separately from the land this water irrigated"—users who "believed they had absolute ownership of the water they used" and who let water they could not sell "flow into marshes in spite of the needs of neighboring farmers" (Trottier 1999: 22; see also El-Eini 1996; Rouyer 1996). The irrigation that did exist consisted of the age-old methods of "aqueducts that conveyed spring water by gravitation and cisterns that collected rainwater" (Karlinsky 1983: 118) and localized irrigation by privately held or owned wells.

The commission located the problem in the legacy of Ottoman law and the "undefined nature of water rights" (Trottier 1999: 21). The Ottoman principle was that surface water that did not flow entirely within private land belonged to the state. However, this principle had not been backed by regulation that would have implemented it, and the incremental historical result was that water rules varied from one village to another, with each relying on local customary laws and traditional norms. It was this decentered system that Jewish private farming had been joining since the late nineteenth century, incrementally gaining control over local wells through the acquisition of land.

The Hope-Simpson Report advised the government to assert its sovereignty and to create a centralized water regime. On the scientific side, it advised an all-Palestine hydrographic survey. On the legal side, it urged government control of all available water sources and strict regulation of irrigation. Relying on this advice and on colonial experience elsewhere, the government later established a central Water Board and drafted its first irrigation bill. Zionist institutions opposed these initiatives, resisting both legislation and state-sponsored scientific studies that could undermine their own scientific claims about the abundance of land and water still available in Palestine for Jewish immigration and, moreover, that would deprive them of their control over the wells and water sources they already controlled and managed (Trottier 1999; El-Eini 2006).[40] This hydro-politics, finally, became all the more acute with the rapid advance of Jewish high-tech agriculture fed by electricity and the advantages of science, capital, and self-organization.

Citrus Farming: The Second Half of the Jewish Economy

The wave of immigration that drove the attachment of the grid to industry also drove its attachment to agriculture. While most newly arrived Jewish immigrants settled in Tel Aviv, many turned to private farming, relying on citrus cultivation as a potential source of a new prosperity. Incrementally, from 1925 onward, citrus products became the largest single item among the country's exports.[41] New Jewish settlements were established all along the coastal plain between Tel Aviv and Haifa, joining the earlier Jewish agricultural colonies and the even older Arab citrus-farming towns and villages.[42] Citrus farming was water intensive, but there were no large-scale irrigation schemes to feed it. Traditional methods of citrus farming depended on animal power and later on steam- and oil-powered engines for pumping water from local wells that catered to small areas.

With electricity in principle available in the District of Jaffa from 1923 onward, the transition to electric pumps was at first slow and selective. Small-scale and localized irrigation technologies that relied on wired electricity were deemed to be expensive in terms of both maintenance and initial capital investment. Electric pumps also had to compete with widely used oil-fueled internal combustion engines. Beyond issues of cost, the older technologies enjoyed the advantage of established habits: an already developed technical infrastructure of maintenance and repair and a general familiarity with the advantages and shortcomings of the machinery. From the perspective of the grid, then, every new connection to an electric pump counted dearly. The prospects of adding more such connections seemed promising, at least as long as private wells were the primary source of water for orange groves.

The citrus farming that was developed by the Jewish immigrants of the mid-1920s was based on the idea that it had to be modernized. Agriculture was to become an industry, and citrus farming was to become a science (Elazari-Volkani 1925). Orange groves now demanded reliance on modern machinery and cutting-edge techniques, as well as reliance on the systematic production of statistical data and scientific know-how. Agronomists focused on issues such as the optimum size of orchards and distance among trees; the amount and timing of irrigation; and the standardization of planting, cultivation, and classification of product quality. Also, hydrologists began to prepare a sound basis for identifying suitable wells and designing water-pumping techniques (Gross 1987). Scientific books, periodicals, and manuals, comparative statistics, and farming conventions became an inherent part of the trade. The Hope-Simpson Report gave a sharp account of the result: "There is no doubt that the pitch of perfection to which the technique of plantation and cultivation of the orange and grapefruit have been brought in Palestine is due to the scientific methods of the Jewish agriculturist" (1930: 94). The "planter-capitalists" (Shalev 1989: 96)—rapidly settling all along the coastal plain—adopted the so-called California model of mechanization, intensive private capital investment, cultivation of larger-scale orchards, and new irrigation technologies (Karlinsky 2000).

The grid also expanded. A new high-tension line was completed in the summer of 1927, stretching from Haifa to the agricultural settlements and orange groves on the coast; another high-tension line started to climb its way north from Tel Aviv, eventually meeting the Haifa line in the early 1930s. The colony of Pardes-Hanna ("Hanna's Orange Grove")—roughly midway between Haifa and Tel Aviv—was established in 1929 and a year later had wired electricity to its wells. The colony of Degania—in the Upper Jordan Valley—received its electricity from the small diesel-fueled station in Tiberias and had 800 dunam (out of a total of 7,000) under irrigation by electric pumps (Hope-Simpson 1930).

The attachment of electricity to orange groves was slower in the older, prewar Jewish colonies and among Arab farmers. In 1930, 90 percent of the new Jewish settlements but only 31 percent of the older ones relied on wired electricity for irrigation (Karlinsky 2001: 167–168). Still, the connection of the grid to pumps and wells made steady progress in the older Jewish colonies as well. These colonies, however recalcitrant some of their farmers, were on the grid by the end of 1926 and early 1927, and at least some of their orange groves already depended on wired electricity for irrigation.

By comparison, Arab attachments to electricity for irrigation were scattered and almost negligible.[43] Jewish settlers enjoyed significant advantages in terms of capital and financing, and were able to pool resources and cooperate on a collective basis (Nadan 2006). Although mainly consisting of private individual landowners, Jewish colonies organized to create "water institutes" that included the well, a water reservoir, and electric "drilling pumps" that, in turn, provided both irrigation and water supply for individual users. The required capital investment, according to then prevailing calculations, could pay off only for irrigating at least 120 to 150 dunam of orange groves. For instance, the agricultural settlement of Ramat Ha-Sharon (established in 1923) made such an investment: its new water institute cost 6,000 Palestinian pounds and began to operate in the summer of 1929. It relied on an electric engine that fed two centrifugal pumps, a 42-meter deep well, a water reservoir with the capacity of 700 cubic meters, and 600 meters of water pipes. The institute provided water for many small groves, some of which were 12 to 24 dunam, that could now rely on the power of collective sharing (Karlinsky 2000: 170).

Arab farmers typically lacked access to the investment capital and cheap credit needed for the shift to wired electricity. Absent this shift, the capacity to organize and cooperate in the standardization and optimization of citrus farming was more limited. The Hope-Simpson Report insinuated a "problem of reasoning": "The Jewish growers are taking steps to ensure by cooperation that the standard of the fruit dispatched from their groves is uniform and that grading and packing are satisfactory. It is regrettable that the Arab growers are not yet convinced of the necessity of the adoption of similar measures" (1930: 94). Comparisons showed that the average cost of maintaining an Arab orange grove amounted to only 60 percent of the cost of maintaining a Jewish one (Metzer and Kaplan 1990: 39). A clear distinction could be drawn: a capital-intensive Jewish agriculture and a (cheap) labor-intensive Arab one.

The Hope-Simpson Report made official the ethno-national difference that developed in Palestine in the second half of the 1920s: "The condition of the Arab fellah [farmer] is little if at all superior to what it was under the Turkish regime," while "the Jewish settlers have had every advantage that capital, science and organization could give them" (1930: 141). And science played its part not only in facilitating the advantages of Jewish farming but also in affirming and consolidating agriculture as another part of an emergent Jewish economy, distinct from the national economy and superior to its Arab counterpart.[44]

Just as in the case of differential supply of "light," access to capital and organizing capacities of Jewish private agriculture—a kind of "organized capitalism"—made the difference between the reach of the grid to Jewish farmers and its inaccessibility to Arab ones. And just as in the case of industry, the attachment of the grid to electric pumps and the network of kilowatt-hour measurement that affirmed the primacy of the category of "irrigation" told an almost exclusively Jewish story. However, also as in the case of industry, the connections between electricity and Jewish contact points—both industrial and agricultural—were neither ideological nor political. The Electric Company was not so vehement in defending its water rights in the Auja against the Mandatory State as a way to protect Zionist interests or Jewish agriculture. In fact, it may have run against the interests of farmers to lure them into dependence on costly electric pumps while idly sitting on the water reservoirs of the Auja.[45]

Concerns over the ability to establish more connections and to yield more profits drove the attachment of the grid to the pumps of hundreds of scattered local wells. Yet the effect of this model of isolated attachments was that the formally universal grid actively took part in producing the growing divide between Arab and Jewish agriculture in Palestine. The outcome of the process of electrification, unlike its original drivers, had ethno-national implications. The physical deployment of the wires and the attachment of the grid to the circulation of capital, rather than any Zionist master plan, participated in the assembly of a politics of separatism. The prospect of separation between Arabs and Jews, just barely on the horizon in the 1920s, was already circulating through the wires.

Conclusion

USING ACTOR-NETWORK THEORY in historical sociology is different from deploying it through observations in a laboratory or in an administrative office. In historical studies, we cannot observe the actors but only examine documentary traces of texts and other graphic representations that once moved from one place to another before being filed in this or that archive. These documents were once actors in the drama of electrification. Some of them were produced by actors who took part in the electrification process. Some remained silent, only represented by others. Some actors were individuals, some were members of organizational entities, others were human-made objects, and still others were outright nonhuman.

The method used here strived not to discriminate among actors. All were equally relevant. And, perhaps, when reading documents and looking at maps, it was easier not to favor a human over a nonhuman actor. While it was obvious that actors participated in assembling an electric grid, I had to keep in mind that the grid was at once an actor and an actor network. This proved useful during my inquiry. Also useful was the fact that in taking a historical view, the "end" was known: Palestine was electrified, but the nominally universal grid produced the effect of a widening Arab-Jewish divide.

It would be convenient and perhaps even elegant to conclude that the electrification of Palestine in the 1920s reveals the yet understudied process whereby the construction of a material infrastructure acted as a means to achieve British-Zionist colonial ends. After all, and notwithstanding some conflicts and disagreements, this process was a product of Anglo-Zionist cooperation. And after all, regardless of formal declarations, the Palestine grid did not promote economic development for everyone. The accumulated evidence shows that electrification benefited the Jewish population more than it served the Arab population, that it sowed crucial seeds of difference between the two communities, that it deepened if not enabled the economic divide between them, and that it was an early precursor to what eventually matured in the 1930s and 1940s into a full-blown episteme of ethno-national separatism. Yet even on such terms, my purpose was not to suggest that the findings speak of any definite causal link between electrification and separatism. Paraphrasing Max Weber, I had no intention of maintaining such a foolish and doctrinaire thesis as that separatism or the divisions and gaps between Jews and Arabs could have arisen only as a result of the grid. I merely proceeded by investigating whether and how certain attachments between electric wires and their newly constituted, demanding consumers produced some observable effects of ethno-national difference.

Considered from the historical standpoint of "knowing the end," such a general account of the roots of separatism is plausible. Only I insisted that the carefully produced, assembled, and networked *effects* of electrification—rather than some overarching forces that pushed it from behind—could explain some material aspects of ethno-national difference. I looked at the spread of wires without a priori assuming that some initial input determined the output. I believe that the findings of this study are all the more telling precisely because they show that the effects of the grid were not planned, preconceived, or "structurally determined."

Then again, it would also be plausible to conclude that what this study shows is that the British and the Zionists found a capitalist who established a commercial enterprise—Pinhas Rutenberg and his company—that embodied the circulation of capital in their actions and were therefore able to bleach out the political project of securing Zionist interests at the expense of Arab ones. To use the old distinction of Henry

Maine, it may be shown that the grid produced its attachments by "contract" rather than through "status" but that these contracts in fact served to conceal superior Jewish financial and organizational capacities that had their origin elsewhere (i.e., unrelated to electrification). My findings provide some solid evidence in support of this conclusion, especially with respect to the attachment of the grid to Jewish industrial and agricultural demand. I could have made this argument, but then I would have remained in the domain of bigness—colonialism, government, capitalism, and nationalism—in light of which I could have framed the process of electrification. I would have stayed with macro-level patterns, offering an empirical study of micro-level interactions and hoping to produce a more or less plausible account of the relationship between the two.

This would not have been a bad choice if I had decided to consider processes of electrification from the perspective of the interpersonal or even interorganizational networks that enable them in any given historical setting. Mark Granovetter moved in this direction in his early work, offering to measure relations among humans in terms of distance and intensity so as to provide "the most fruitful micro-macro bridge" sociology can offer (1973: 1360).

Work on interpersonal networks has offered sociology some valuable insights—but at considerable cost. For one, interpersonal networks have quite quickly become identified with the term "social networks," in this way misleading the social into the golden cage of being an autonomous domain. Social networks can now be studied independently, as if they have an existence that may be analytically distinguished from all that takes place around them. The downside of this approach is that it pushes sociology into being a science of the residuals, assigned to those inquiries that remain outside the scope of interest of the real scientists who work on their phenomena in their own jurisdiction (e.g., uncovering the social background of judges who reach an important constitutional decision on the basis of principled legal reasoning). Also, the focus on social networks leaves out all of the nonhuman elements that participate in the movements that sustain networks. Some of these elements are human-made and some are not, but in both cases they disappear from view, treated as sociologically neutral. John Law has a rather vivid way of articulating this problem: "Would I still be a sociologist if you took away not only my colleagues but also my office, my books, and my computer?" (1992: 382). Finally, the idea that interpersonal networks offer a bridge between the micro and the macro is premised on the notion that these are two distinct hemispheres, the former made of everyday small gestures and local exchanges (i.e., action and face-to-face or even face-to-screen encounters); the latter made of patterns, blueprints, and formalized relations (i.e., structure).

In this study, I chose not to discriminate small from big, action from structure, and micro from macro. At least not in advance, at least not before the inquiry began, and definitely not by prioritizing one over the other so as to end up studying a concrete local process that securely took place "in context." Bruno Latour (2005) describes such a theoretically guided methodological decision as an attempt to flatten the territory of inquiry, to unfold the folds in the terrain. In such an inquiry, the Government of Palestine and an electric meter, the Municipality of Jaffa and an electric water pump, or a memorandum circulating the Colonial Office, and kilowatt-hour measurements of industrial demand all assume equal standing. However, this method is not merely an attempt to enrich social networks by fusing them into technical ones, or to offer a more elaborate bridge between macro-level patterns and micro-level interactions. What the method tries is to trace the assembly of the social without presuming that a micro and a macro (or action and structure) are already there as its constitutive elements.

So, then, at the core of the analysis and the findings of this book lies an attempt to address two basic questions: What is an actor network? And what is the difference between an actor network and actor-network theory? Throughout this study, I have treated an actor network as a more or less stable set of connections between heterogeneous elements. The more stable it is, the less visible its connections, elements, and circulations. In its most stable form, the actor network disappears from view altogether and reappears as a single coherent source of action; it appears to behave as if it is only an actor.

A common sociological way of describing this level of stability is through the concept of institutionalization, in which some procedures for maintaining connections are fixed, some patterns of circulation are solidified, and some movements are not disrupted. This is the reason that actor-network theory gained its original fame in the study of scientific and technological innovations: these are situations in which the actor network is most visible (the same is true for accidents and breakdowns). Issues and matters are not yet fixed. Procedures must be tested, routines have to be fought for, surprises are expected, the single actor (or phenomenon, force, cause, source, category, etc.) is still to be produced, and hence it is easier to consider what is going on as an effect of an emergent network. An actor network is real in the sense that it is possible to identify and locate it by empirical means. If one is curious to do so, of course.

Actor-network theory is mainly about teaching sociologists how to become actor-network painters. It equips them with some incentives to see why it may be worthwhile to become painters, essentially pushing the case that it may be useful and beneficial to the sociological craft. The how and the why are intertwined, and this is why actor-network theory is famously (or notoriously) recognized in its "material-semiotic" empirical studies. It is radical in its nonessentialist and nonreductionist approach: "The social is nothing other than patterned networks of heterogeneous materials" (Law 1992: 381). Another premise follows: the fundamental units of sociological inquiry—matters of concern such as power, inequality, groups and organizations, subjectivities, institutions, distinctions and categories—are all effects of actor networks. So the task of the sociologist is to identify actor networks and to use a pen or a keyboard to draw them and detect their effects. This is no easy task because actor networks cannot be mapped like "social," "organizational," or "technical" networks.

Actor networks "can only reveal themselves when activated" (Buzelin 2005: 197). They are uncertain and unpredictable in the sense that their failure to become is as viable an option as their success; they are about creativity: about making some connections and hoping they hold. So there is a basic difficulty here for the actor-network painter (the sociologist with her colleagues and books and keyboard) because the painting to be is not an ordinary one. It tries to capture a performance without fixing it; it tries to show how a certain connection is established but most likely disappears from view as soon as it is made, either because it now acts as a single actor ("successful" in being an entity, a discovery, a recognized phenomenon, a source of power, etc.) or because nothing happens ("failure"). That is why it is probably more accurate to describe an actor-network painter as one who is equipped with a certain type of brush—one that traces associations and connections—rather than with a definite product we can comfortably designate as a painting.

And here also lies the biggest challenge and problem of actor-network theory: Where does one stop? How does one decide what are the observable units or elements of an actor network when each of these elements may be a (hidden) actor network in itself? One direction of inquiry is to wonder how it is that "we are sometimes able to talk of 'the British government' rather than all the bits and pieces that make it up" (Law 1992: 381). One of the premises of actor-network theory is that "all phenomena are the effect or the product of heterogeneous networks." This means that even individuals, let alone an entity like the British Government, are what they are because they are a "patterned network of heterogeneous materials" (1992: 381). Thus, one way of positing a research question is to ask about the processes or the strategies of patterning and ordering that make actor networks look like actors who are a single-point source of action. And yet from the point of view of method, it is necessary also to turn the question on itself and ask how we can paint an actor network if at least some of our elements are also actor networks. How can we detect and follow an actor network if we know that all of its actors are effects of some other actor networks? Does actor-network theory ask us to be actor-network painters ad infinitum, if not ad absurdum? There is no simple solution to this problem. It is one thing to suspend judgment about big and small or about micro and macro, and it is quite another to ignore the fact that some of the actors in "our" actor network are also an effect of actor networks and to proceed as if they are not.

It seems that solutions can only be partial and, more important, empirical rather than theoretical. They have to do with what we want to find out, with proceeding slowly, and with making compromises. In this study, I wanted to explore whether electrification could tell us something about the ways ethno-national difference is produced and performed. My principled decision was to follow the wires—to accept the electric grid as a visible material assembly whose progression to and through various contact points could be ascertained with a high level of confidence. At least some of the materials that make up electric grids are quite durable. It is a good ordering strategy for any network, writes John Law, "to embody a set of relations in durable materials" (1992: 385) and hence to secure stability of performance. At the same time, "durability is yet another relational effect, not something given in the order of things" (1992: 385). The study of the electric grid proved to be a good candidate for demonstrating this point: on the one hand, a techno-legal network of contracts, wires, and measured units of electric energy that was stable enough to act as a single-point actor; on the other hand, an actor whose integrity and durability depended on relating and attaching itself to new sources of demand.

I collected texts and graphic representations that documented the techno-legal and techno-administrative work involved in the assembly of an electric grid and its gradual expansion. I treated the grid as a purposeful actor while keeping an eye on the various circuits of heterogeneous circulations that participated in its assembly. One such element was the Electric Company. The company was in charge of many related technical and administrative issues and was therefore treated as an actor. It also behaved like a single-point actor when it dealt with other entities such as municipalities and the government. Yet it was also a complex assemblage of bits and pieces (machinery, electricians, maps and tables, shareholders, etc.), and I analyzed some of these bits and pieces as elements of other actor networks. For instance, I found that the water rights that were held by the company had a role in sustaining "modern Jewish agriculture" and that the measured category of "industrial demand" could be attached to a scientifically validated "Jewish economy."

And it is this double gesture, so to speak, that allowed me not to settle for following the wires and poles to their newly constituted electric consumers but to also (and simultaneously) consider how they became elements in other actor networks with effects other than acquiring electricity consumers. It is by allowing the Electric Company, as well as the government and current meters and electric pumps, to mutate that I can suggest with some certainty that inequality between Arabs and Jews, regional differences between them, and an emergent ethno-national dual economy, as much as streetlights, were observable effects of electrification. Consider the current meter. On the one hand, it was an actor that marked the boundary between the general distribution system and private consumers. On the other hand, it was an actor network of kilowatt-hour measurements, statistical categories ("private light," "irrigation," etc.), social classifications ("upper classes," "Jewish immigrants," etc.), and scientific studies, which in turn played a role in the assembly of a Jewish economy.

And then again, some actor networks were also at once actors that participated in the expansion of the grid. So, for example, I traced the role of the Electric Company (and visible electric poles) in constructing the actor network of "Arab agitation" and showed that the latter became an actor that determined technological and legal decisions (e.g., where to locate a powerhouse and whether to extend the Auja Concession). This is the three-dimensional view that actor-network theory allows, preserving and respecting the capacity of materials and elements to mutate between their role as actors and as actor networks, moving back and forth between some silent backstage to a more noisy front stage. And still the decision of when to turn a blind eye to some actor networks and allow them to act as single-point actors remains arbitrary in the sense that it is the researcher's judgment that determines when to allow and when to disallow actor networks and actors to switch roles. There cannot be a prescribed answer to the question "Where does one stop?" but there may be a diagnosis of the final contour of the brush: when the bigness of colonialism, Zionism, and Arab-Palestinian nationalism somewhat dissolved into switching on lights and paying electrical bills.

Notes

Archive Abbreviations

BNA: [British] National Archives, Kew, Richmond, Surrey, UK
CZA: Central Zionist Archives (Jerusalem)
ECA: Israel Electric Company Archive (Haifa)
ISA: Israel State Archives (Jerusalem)
PLA: Archive of the Pinhas Lavon Institute for Labor Movement Research
RLZ: Historical Archive of Rishon Le-Zion
TAHA: Tel Aviv Historical Municipal Archive (Tel Aviv)

Introduction

1. Joel Achenbach, in his article "The 21st Century Grid," *National Geographic* (July 2010), describes our critical attachment to the electric grid as that "endless supply of gadgets that demand to be fed on current. We march around with chargers at our disposal for laptops and cell phones . . . [and] when our devices die, we become restless for power, expressing disbelief that power is denied to us." Retrieved March 3, 2012, from http://ngm.nationalgeographic.com/2010/07/power-grid/achenbach-text. About the recent crippling effects of failing grids, see, for example, July 7, 2012, news stories about the hundreds of millions who were left without electric power in northern India, the loss of which also led to the failure of the transport system; also see the reporting on Hurricane Sandy's effects on electric and water supply in New York and New Jersey in the fall of 2012.

2. On April 3, 2012, the *Jerusalem Post* quoted Israel's Minister of Energy: "For Israel it is also important to just be connected to Europe. . . . We are an energy island and this has its political significance." Retrieved May 2, 2012, from http://www.jpost.com/DiplomacyAndPolitics/Article.aspx?id=260410.

3. The "war of currents" refers to the eventual triumph of alternating current, suitable for stable long-distance transmission, over Edison's direct current (see Hughes 1983: 106–109).

4. In this, electric current somewhat resembles the behavior of what Latour calls "an immutable mobile" (Latour 1987: 226): a model or object that does not undergo

transformation while moving from one location to another (and hence plays a crucial role in enabling and sustaining networks; see Mol and Law 1994). This "property" of alternating current, precisely because of the nature of its flow, must nonetheless be qualified: the technology of circulating electric current also includes as a key component the existence of "transformers stations" (big and small) for transforming the "tension" (voltage) of current from higher to lower. In this sense, electric current is a hypermutable mobile, albeit controlled and predictable such that a level of consistency similar to that of the immutable mobile is retained.

5. For a user-centered approach, also see Wilhite et al. (1996) and Winther (2008a). Relying on this approach, some researchers have begun to pursue the "anthropology of energy" (Wilhite 2005). Wilhite et al. (1996) compare the cultural and economic factors influencing the use of electric heating, lighting, and hot water in Norway and Japan. Tanja Winther, noting that "relatively little has been done to understand people's use of energy" (2008: 2), has studied the impact of electrification in Zanzibar on gender relations and family life, home economics, and time management (also see Chambers 1994). For a comparative study of how electricity affects rural communities, see Barnes (1988).

6. See Furlong (2010) on theoretical approaches to infrastructures, particularly her claim that they are less stable than they look.

7. Gilbert writes that the British hoped that their promise to facilitate a Jewish National Home would encourage Russian Jewry to persuade Russia to stay in the war, would stimulate American Jewry "to accelerate the military participation of the United States," and later would possibly secure the support of Russian Jews for Zionism at the expense of Bolshevik sentiments (2007: 27–28, 37–38).

8. Rutenberg occupies a unique and controversial position in the history of Zionism. An outspoken capitalist and yet a figure with some political aspirations, he often entered into bitter personal and principled conflicts with the Zionist establishment. Eli Shaltiel's excellent political biography of Rutenberg clearly shows that it would be wrong to account for his activities in matters related to the electrification of Palestine in terms of Zionist "interests." As a political biography, Shaltiel's account says little about the Auja Concession and in general does not focus on the actual electrification process. Sternhell, in passing, describes Rutenberg as an antiunion charismatic and authoritarian capitalist who nonetheless enjoyed warm relations with the labor movement because of a shared preference for the "practical" (1995: 301–302).

9. Concessions are by no means only a colonialist prerogative. Sovereign governments, seeking investment in exchange for "development" also employ them. Veeser classifies concessions as five major types: territorial (more like treaties: for example, the Suez and Panama canals and the vast concessions in China), colonial, speculative, discretionary, and neoliberal. He suggests that "concessions are an institutional foundation of modern capitalism about which we have virtually no systematic, comparative knowledge," rendering them "uncollected, under-analysed and unquantifiable" (Veeser, forthcoming).

10. The British were not alone in the use of concessions. Colonial concessions were widespread in resource-rich Africa and were granted by all major colonial powers. Most notorious among these were concessions granted by Belgian King Leopold to private companies in the Belgian Congo, where atrocities and forced labor were employed in order to benefit from the "rubber boom" of the 1890s (Ascherson 1963; Ewans 2002). Concessions for extracting minerals, diamonds, gold, and copper; for using land; and for constructing railways and ports were granted to private companies by the French, Portuguese, and Germans as well. "By 1914," writes Veeser, "concessionary companies controlled one-third of German South-West Africa" (forthcoming). The British granted thousands of concessions for mining, rubber, and palm oil in its colonies. At times, chartered companies were granted exclusive privileges on a territorial basis, allowing them to invest and profit by a variety of methods in more than one area of commercial activity. For example, "Goldie's consortium was chartered by the British government as the Royal Niger Company and granted broad concessionary powers in 'all the territory of the basin of the Niger'" (Ikein et al. 2008: 3). Deeper into the twentieth century, concessions were widely sought and granted for oil exploration and extraction in Africa and the Middle East.

11. The political processes and circumstances surrounding the grant of concessions to Rutenberg are thoroughly analyzed by Shaltiel (1990) and are not discussed at any depth in this book (also see Smith 1993; Reguer 1995).

12. BNA CO 733/3, May 1, 1921, dispatch no. 80 from Herbert Samuel to Churchill.

13. BNA 733/81, July 29, 1924, file of legal opinions on the Mavromatis case, and September 8, 1924, minutes following the decision.

14. See Greece v. Britain, Permanent Court of International Justice, File E. c. V., Docket VI 2, Judgment No. 5, March 26, 1925, and Greece v. Britain, Permanent Court of International Justice, File E. c. XII, Docket XII 4, Judgment No. 10, October 10, 1927. Also see BNA CO 733/80 and CO 733/131/3 for full details on the court proceedings and accompanying legal briefs. As a follow-up, the Government of Palestine formally recognized the Mavromatis Concession; see BNA CO 733/119, February 25, 1926, "Concession to Mavromatis as to Electricity in Jerusalem," proceedings and accompanying legal briefs. The Mavromatis cases became widely known and are still considered part of the canon in international law. For an early analysis, see Borchard (1925).

15. Both companies were financed by Zionist organizations and philanthropic sympathizers. They were effectively managed and controlled by Rutenberg, Baharaw (his secretary), Sacher (his lawyer), and Berkenheim (his chief engineer).

16. Ben-Eliezer (1995) shows that this self-confidence also relied on a Jewish military build-up from 1936 onward, actively preparing for a military solution to the "Jewish-Arab problem."

17. The controversies among Zionists over the 1937 British Partition Plan have been studied by Dothan, who writes, "Prior to 1937, the question of Jewish sovereignty over Eretz-Israel [Palestine] did not occupy a major place in Zionist thought" (1979: v).

18. The issue of the Palestinian Nakba (the "catastrophe" of dispossession and expulsion) is still debated and contested, but it now occupies center stage in Palestinian historiography. See, for example, Kimmerling and Migdal (2003), Masalha (2005), Sa'di and Abu-Lughod (2007), Lentin (2008), and Kassem (2011).

19. "Yishuv" is the most common term used for marking the distinct characteristics of the Jewish community in Palestine. A shorthand for the "organized Jewish community of Palestine," it also appears alongside phrases such as "state within a state" and "on the road to statehood." All in all, the terminology is one of separatism and ethno-national and ethno-political distinction; see, for example, Horowitz and Lissak (1978).

20. There are also histories of the Arab-Palestinian population—although considerably fewer than those of the Jewish population—that analyze the ruptured composition of Palestinian society and the troubled political dynamic of the Palestinian national movement (Khalidi 2006, 2007; Khalaf 1991).

21. Much scholarly research seems to have followed the results of the Zionist World Organization's 1920 London conference on this issue. This conference staged the famous Weizmann-Brandeis debate on how to further advance the Zionist cause: Should it rely on investment and capitalist methods of development or on collectively organized and dispensed public capital? Louis Brandeis, representing American Jewry and an American capitalist spirit of private initiative, "lost the debate." See Shapiro (1973) and Halpern (1987). For an Arab-Palestinian analysis of capitalism in Palestine, see Zu'bi (1984).

22. The dual-society paradigm was articulated in the Palestine Royal Commission Report to the Secretary of State for the Colonies (the Peel Report) in 1937. The sociological analysis of Horowitz and Lissak bears a strong similarity to the sociopolitical analysis offered by the Peel Report 40 years earlier; see Shamir (2000:18–19 and n. 33). Studies that are critical of Zionist ideology, written from the perspective of colonial analysis and neo-Marxism, tend to share the self-reliant perspective. For instance, see Shafir's emphasis on the "creative pragmatism" of Jewish settlers (1993: 109); also see Shalev (1989), who analyzes the role of organized Jewish labor in espousing Jewish separatism in the labor market. However, such studies deviate from the Zionist norm by emphasizing the exploitation and exclusion of Arabs. See Zureik (1979).

23. Quoted in BNA PREM 1/24, June 22, 1923, "The Palestinian Arab Case," prepared by the Palestinian second delegation to London.

Chapter 1

1. One dunam equals 1,000 square meters equals 1/4 acre equals 0.000386 square mile.

2. BNA CAOG 14/109, The Auja Concession.

3. Yaari-Polskin (1939: 252), Naor (2003: 36–37, 42), and Smith (1993: 121). I could find only one account that doubted the "Arab agitation" explanation. Historian Shmuel

Avitzur suggests two other motives for relocation: one, the concern that the conflicting claims of Mavromatis over the right to exploit the waters of the Auja River would stall Rutenberg's plans; two, the realization that the water power of the Auja would not generate sufficient energy. See Avitsur (1957: 146).

4. Siegfried (Eliezer) Hoofien, an influential Dutch-born banker, acted both as general manager of the Anglo-Palestine bank that financed the Jaffa Electric Company and as a director of the company on behalf of the Jewish Colonial Trust (a large shareholder). He was well known to the Arab leadership as a major Zionist figure. In 1923, the Electric Company negotiated with the Municipality of Jaffa on terms for the supply of streetlights; Rutenberg gave explicit instructions to have representatives of the company other than Hoofien sign the contract so as not to provoke Arab officials (ECA 352-111). On other, less politically sensitive, occasions, Hoofien was often a signatory on behalf of the company (see, for example, CO 733/151/6).

5. For full lists of shareholders and their voting rights, see ECA A-0475-20, A-0475-55, and A-0475-90. For an in-depth analysis of the financial backing of Palestine electrification, see Shaltiel (1990).

6. In 1922 and again in 1923, Arab delegations to London tried to have the official British policy of facilitating a "Jewish National Home" in Palestine rescinded, and in the process raised the matter of concessions as an example of unfair preferential treatment of Jews and as a "symbol of Zionist aggression" (Smith 1993: 121). See BNA CO 537/854 and PREM 1/24, both of which relate to the 1923 delegation of the Fifth Palestine Arab Congress, led by Moussa Kazim El-Husseini Pasha. The delegation claimed that "the concessions promised to Mr. Rutenberg, one of the most unjust ever given, indicates [*sic*] the way in which Palestine is being administered on behalf of the Zionists. It secures to the Rutenberg Company, which will be composed of Jewish financiers, a stranglehold on the economic life of Palestine and Transjordania for 70 years, with a chance of further extension. It was not put out to public tender, an indication that it was meant not only for Economic but also for Political purpose." See BNA PREM 1/24, June 22, 1923, "The Palestinian Arab Case."

7. ECA 2371-1-43, July 13, 1923, unsigned and unaddressed letter carrying a handwritten note, "handed to Mr. Vernon" (Vernon was an official in the Colonial Office, closely involved with matters relating to Rutenberg's concessions). Also see ECA 2371-1-64, July 24, 1922, letter from Rutenberg to the Undersecretary of State for the Colonies.

8. ECA 2371-1-79, January 2, 1922, letter from Sacher to Rutenberg.

9. One Palestinian pound (LP) equals 1 pound sterling (£) equals 1,000 mills equal 97.5 Egyptian piastres (PT) or 975 milliemes (mm). An Egyptian pound equals 1,000 mm or 100 PT equal LP 1.02564.

10. The general circumstances and concrete negotiations that eventually matured into the Auja Concession and the Jordan Concession, and the controversies they sparked in London, have been thoroughly studied and reported on by a number of historians. These

historians mostly consider the matter in light of the "political context" of Zionism and Jewish-Arab relations, specifically the Balfour Declaration that committed the British government to a Jewish National Homeland in Palestine. Some also note the unique personal relationships and social ties between Rutenberg and British and Jewish British politicians and financiers. For such in-depth discussions see, for example, Shaltiel (1990), Huneidi (2001), Yaari-Polskin (1939), Naor (2003), and Smith (1994). A more expansive view is offered by Reguer (1995), who considers the granting of concessions in light of broader imperial interests such as the railways of the Near East and other strategic and financial considerations (see Chapter 4).

11. ECA 2371-1-64.

12. ECA 2371-1-59, November 22, 1922, letter from Rutenberg to the Colonial Office.

13. Ibid.

14. ECA 2371-1-46, June 11, 1923, letter from Rutenberg to the High Commissioner for Palestine.

15. ECA 2371-1-43, July 13, 1923, unsigned, unaddressed document, which could not be located in any other government archive in Jerusalem or London.

16. ECA 352-111.

17. BNA CAOG 14/108, December 23, 1921, report submitted by Preece, Cardew & Rider to the Colonial Office. These technical specifications were inserted almost verbatim into the concession agreement (articles 8 and 9). Also see ECA A357-195.

18. This Westminster-based London firm was formerly called Preece, Cardew & Snell. The late Sir John Snell, a one-time president of the Institute of Electrical Engineers, was a senior partner from 1910 to 1921. On the history of this firm, see Baker (1980).

19. BNA CAOG 14/109, December 23, 1921, report submitted by Preece, Cardew & Rider to the Colonial Office (hereafter Preece Report) BNA CAOG 14/109, February 2, 1922, follow-up letter from Shuckburgh to Rutenberg (on proposals by the consulting engineers for the electrification of the Jaffa–Jerusalem Railway and the Hedjaz Railway) (also see Chapter 4).

20. BNA CAOG 14/108.

21. BNA CAOG 14/109, letters from John Shuckburgh to Burchells, solicitors to the Colonial Office.

22. See BNA CO 267/627/1-8 on the "damaging" role of Preece Cardew & Rider in the matter of the Sierra Leone power station breakdown.

23. BNA CAOG 14/108, December 23, 1921, Preece Report.

24. BNA CAOG 14/109, August 9, 1921, memorandum from Preece, Cardew & Rider to Vernon at the Colonial Office: "The Auja basin concession was, as you know, not referred to the consulting engineers. . . . The terms of the Auja basin concession were practically agreed by you with Mr. Rutenberg before referring to us and do not admit of revision."

25. BNA CAOG 14/108, Preece Report.

26. Ibid., p. 6, para. 9.

27. The recommendation to include such a clause in the concession also corresponded with the general legal instruction to prepare the concession "in the form used in Colonies and Protectorates." Attempts to devise and oversee a "colonial model concession" did not seem to work very well. "Model concessions" underwent constant changes, adapting to local conditions. The concession network may therefore be described as a failed one (Mol and Law 1994).

28. The report also mentioned the fact that the firm's deceased partner, Sir John Snell, had had a previous opportunity to favorably comment on Rutenberg's plans. It seems to have referred to a letter sent by Snell to Mond on August 10, 1920, praising Rutenberg's qualities and expressing support for the hydroelectric scheme. Sir Alfred Mond (Lord Melchet) was a highly influential British industrialist (and Jewish Zionist) who was part of the close circle supporting Rutenberg's electrification plans. He was soon to become a director on the Palestine Electric Company's board. The Jaffa Electric Company was absorbed into the Palestine Electric Company in November 29, 1927. Also see BNA CO 733/13, March 7, 1921, letter from Snell to Herbert Samuel, cited in Reguer (1995: 699).

29. BNA CAOG 14/108, Churchill's response to telegram no. 342, August 25, 1921, from High Commissioner Samuel.

30. ECA 357-197-001, May 1923, "Transfer Agreement" specifying the electric appliances owned by the Council of Tel Aviv prior to their transfer to the Jaffa Electric Company.

31. TAHA, 283b (box 422), April 18, 1921, protocol 23, minutes of the Council of Tel Aviv Township. Dizengoff reported on Jewish businessmen from Poland who had arrived in Tel Aviv with the capital for establishing small factories but without enough resources for their own electrical generators.

32. Ibid., April 10, 1921, protocol 21, 233–241.

33. Ibid., April 14, 1921, protocol 22.

34. TAHA, 283d (box 422), February 5, 1922, protocol 77, 564–567.

35. An indication of German commercial might can be found in the annual statistics of the Department of Customs, Excise and Trade of the Government of Palestine. In reports for both 1923 and 1924, the director of Customs and Trade observed that imports from Germany were high because "German exporters pursue a vigorous trading policy, favored by generous credits." Statistics showed that Palestine imported practically all of its electric batteries, lamps, motor ploughs, and electric accessories from Germany, leaving British products only a meager share of this business. A similar disproportion held with construction materials such as iron bars, cables, steel, bricks, and cement, with German imports often overwhelming British imports by a margin of fifty to one. See BNA CO 814/1/5 and CO 814/1/13.

36. ECA 2371-1-85.

37. The document compared the price of turbines, not that of diesel engines. At least for public purposes, Rutenberg had to show preparations for the construction of a hydroelectric rather than a diesel-powered powerhouse. However, there is strong evidence to

suggest that he was already preparing for the construction of a diesel-fueled powerhouse at this early stage, prior to the official granting of the concession.

38. Sir Joynson-Hicks, member of Parliament, claimed that throughout his years of experience with colonial concessions he had never seen such irregular ones as those granted to Rutenberg, specifically referring to the absence of a clause requiring the concessionaire to purchase British products and equipment. See Yaari-Polskin (1939: 241), citing the parliamentary debate of July 4, 1922.

39. ECA 2371-1/0086, August 17, 1921, letter from Rutenberg to Weizmann. This letter sheds new light on the belief that Rutenberg sought Weizmann's instructions on the matter but did not receive a response (see Naor 2003: 44). For Weizmann's views on Rutenberg see Weizmann (1949: 212–216).

40. ECA 2371-1/0043, December 21, 1921.

41. Accounts of the scandal are given in Naor (2003: 44, n. 59, n. 60) and Yaari-Polskin (1939: 234–237). Also see BNA CAOG 14/109.

42. BNA CO 323/935, March 17, 1925, Colonial Office memorandum. The price of copper wires in England provides further evidence of German competitiveness. The memorandum complained about British members of the Copper Wire Manufacturers Association: "When tendering to this office, [they] quote prices which are materially higher than those which they quote to other purchasers upon the same specification, and it is alleged that the difference is due to the fact that we do not include Continental manufacturers in the list of firms invited to tender." Tacitly referring to German copper wire manufacturers, the memorandum sought to reformulate the policy so as to allow the Colonial Office, when it shopped for copper wires in the colonies, "to include others if the secretary of state agrees."

43. ECA 0352-111-247, September–October 1923, Rutenberg directives.

44. BNA CO 814/1, "1925 Annual Report of the Department of Public Works," sec. xiv, "Public Utilities Concessions."

45. ECA 358-210-17, December 31, 1925, "Report of the Jaffa Electric Company for Second Year of Operation 1924/1925." Originally a Glasgow firm, Mirrlees specialized in the design and manufacture of a wide variety of diesel engines, including large engines for electric power stations. Its products "were installed in many large diesel generating stations. Many multi-engine contracts were obtained for this engine, both at home and abroad." Retrieved March 3, 2013, from the Anson Engine Museum, http://www.enginemuseum.org/mrindex.html.

46. BNA CO 814/1; also ECA 358-210-17. But see ECA 358-210, 1927 report submitted by Rutenberg to Lord Reading: "The supply of energy could not be increased in view of the fact that the new Fraser & Chalmers set has not been in working order. The new flywheel [a balancing circular mass on the output of the engine crankshaft] was received only a few weeks ago and installed on the 24th [of] March. This engine is of good quality and will, I expect, give full satisfaction" (p. 910). The timing of the purchase of this large engine by the Jaffa Electric Company coincided with the arrival of a senior General

Electric engineer, who came to examine possible investments in the electric works in Palestine; see CZA S25/7416-72, December 11, 1925, confidential letter from Chaim Weizmann to Colonel Fredrik Kisch, Palestine Zionist Executive, Jerusalem.

Chapter 2

1. ECA 0352-111-251, April–May 1923, Rutenberg directives.

2. TAHA 6-54, box 1585.

3. Ibid., January 16, March 25, March 29, July 18, September 9, and September 30, 1923, unsigned letters.

4. Nye writes that electrification in the United States and Europe first developed on a local basis not only because of the limits of direct current but also because "the high price of copper wires made long lines too expensive"(1990: 5).

5. ECA 358-210-018, "Jaffa Electric Company First Annual Report to the Board of Directors, 1923/24," p. 6, article 7.

6. Article 10 of the Auja Concession stipulated: "The concessionaire shall have the right to erect poles for the high medium and low transmission lines and to lay an underground cable or cables through the streets wherever required but so that the concessionaire shall be under the obligation to restore such streets to their former state." The concession did not specify who should bear the costs.

7. Two theoretical approaches in "science and technology studies"—the system approach (Hughes 1983) and actor-network theory (Latour 2005)—nonetheless share the view that we must not separate technical and political aspects in the analysis of material infrastructures and technological and scientific innovations.

8. TAHA 6-54, box 1585, October 7, 1923, letter from the Jaffa Electric Company general secretary to the Township of Tel Aviv.

9. ECA 352-111-244, August 15, 1924, telegram, "Instructions to Jaffa."

10. ECA 352-102-17.

11. ECA 352-111-244.

12. ECA 352-111-247, early 1923, "Jaffa Instructions."

13. See, for example, EAC 352-111-245, August 14, 1924, cable from Alexandria to Rutenberg: "Instructions to Palestine Electric Company," and ECA 352-111-151, ca. 1923, instructions to Mr. Itin (in Hebrew). Also see ECA 475-49, June 14, 1925, "Second Annual Report to the Board of Directors." This report details the purchase of lands in Haifa, Tiberias, and the Jordan Valley.

14. Hughes credits the term "electric grid" to the British Weir committee and specifically to Charles Mertz (1983: 352–354).

15. In 1925, only 4 percent of U.S. farms had electricity. The percentage rose to 10 in the late 1920s, half of which generated by "isolated" electric plants and a considerable portion of the other half based on feeding off "interurban" electric railway lines (Nye 1990: 292–296).

16. "The post–World War I rationalization for electric supply took many forms," writes Hughes. Sometimes they knitted together utilities that evolved independently; sometimes a separate corporate entity "owned and managed the grid"; and sometimes grids "were analogous to a confederation of nation-states" (1983: 324–325).

17. Tinker also studied the situation in Burma. In 1923, only six towns had electricity; in 1937, all fifty-eight municipalities were electrified by private companies (1968 [1954]: 233).

18. TAHA 2-34b, January 23, 1923, letter from Berkenheim to Dizengoff.

19. BNA, MRI-1834, 1922 railway maps.

20. The Government of Palestine recognized Tel Aviv as a 'township' in June 1921, thus sealing in law a place already celebrated as "the first Hebrew city." Township status signaled some administrative and juridical autonomy from Jaffa without setting a definite border between the two. Tel Aviv remained under the jurisdiction of the Jaffa Municipality Town Planning Committee well into the 1930s, but the process of administrative separation was already concluded in 1927 "when Tel Aviv and Jaffa were declared two separate municipal authorities" (Shoham 2012:41).

21. Ethno-national tensions surrounding Jewish immigration resulted in violent clashes in May 1921, but these events did not have a lasting impact on the relationship between Jews and Arabs in Jaffa and Tel Aviv. The 1920s—specifically the first crucial years of electrification (1923–1928)—were peaceful until the eruption of ethno-national violence in August 1929.

22. See Chelouche (2005 [1931], 324–325).

23. TAHA 3-40.

24. TAHA 3-40, March 9, 1925, letter from the Electric Company in response to March 4, 1925, letter from the Township of Tel Aviv.

25. TAHA file 283c, box 422, November 16, 1921, protocol no. 60.

26. A similar phenomenon seems to have characterized the early years of cable television and later the Internet, when inexperienced regulators, finding it hard to distinguish between the physical infrastructure and "content," assigned greater importance to the former.

27. ECA 349-42, June 1922, correspondence between Omar Beitar, chairman of the Jaffa branch of the Moslem-Christian Association, and Asim Said, mayor of Jaffa. Also see CZA KH1-146-2.

28. See ECA 2379-04, November 29, 1922, transfer of property from Rutenberg to the Jaffa Electric Company, including a 20-dunam kushan (registered land deed), no. 1060/22.

29. TAHA 1-283, October 3, 1921, protocol no. 53.

30. TAHA 2-34A, 1921, otherwise undated agreement between the Township of Tel Aviv and Rutenberg. The essentials of the agreement were (1) the rate of electricity to consumers to include a surplus to the municipality (basically 4 percent of gross income

from private residential consumers and 0.5 percent from industrial consumers); (2) the municipality to pay a fixed sum for a basic grid (marked on an attached map); (3) extensions to be made at the expense of the municipality according to special agreements, with the company reserving the right to expand the grid at its own expense in cases where the township did not ask for extensions (and subsequently paying a lower "rate of profit" to the municipality).

31. ECA 357-197-001, May 10, 1923, "Transfer Agreement."

32. TAHA file 283c, box 422, November 16, 1921, protocol no. 60, report from Dizengoff to the Council of Tel Aviv stating that he had met with Jaffa officials and that they were interested in joining the electrification process. Also see December 29, 1921, report from El Akbar in Jaffa, stating that the township approved of the electrification plan and considered it a private commercial enterprise and not a Zionist scheme (Naor 2003: 37).

33. See, for example, Smith (1993: 122), reporting on the Arab press in Jerusalem, which had called for boycotting Rutenberg's plans.

34. See, for example, ECA 349-42, June 1922 letter from Omar Beitar, chairman of the Jaffa branch of the Moslem-Christian Association, to Asim Said, mayor of Jaffa. Beitar wrote that "Patriots" should not approve of the Rutenberg scheme. Also quoted in Naor (2003: 37, 66–67).

35. ECA 0358-210-18, 1924, "First Annual Report of the Board of the Jaffa Electric Company," 5–6. Rutenberg had been vigorously pleading the case of electrification as a commercial business, both to the Arab delegation in London and to Jaffa officials in Palestine. He also acknowledged the "slowdown" effect of political pressures in the first months of the project: "It must be pointed out that during the first six months of supply of electric energy only Tel Aviv and the Jewish inhabitants of Jaffa were provided. The other part of the population consisting mainly of Arabs was boycotting the company. Only after six months of operation the company succeeded in coming to an arrangement with the Jaffa Municipality for street lighting. By that time also applications for light and power from private Arab consumers began to get in. If Arab consumers were to come in from the beginning . . . the income would have been increased."

36. See CZA S25/7416-17, S25/7416-8, S25/7416-5, June 8, 1923, correspondence from Major Campbell, acting governor of the District of Jaffa to the Municipality of Jaffa, declaring the legality of the concession and urging agreements with the Jaffa Electric Company; also see articles in the Arab press criticizing the calls to boycott Rutenberg—for example, *EL Naffir*, April 18, 1923.

37. TAHA 31-73, July 11, 1923, translation of letter to the Municipality of Jaffa.

38. ECA 0352-111-247. The reference to Arab labor is dealt with at length elsewhere in this book.

39. ECA 2371-1 and S25/7416-5, July 13, 1923. Also see Naor (2003: 66) on the acceptance by the mayor of Jaffa of the notion that "the merchants of Jaffa would have to shut

their stores at five in the afternoon while the lights of the shop windows of Tel Aviv will blind the eyes for many more hours." See also Chelouche (2005 [1931]: 407).

40. ECA 0352-111-247, October 1923, Rutenberg's instructions to company engineers and technicians. Rutenberg also advised: "Engage by special recommendation a reliable good Moslem foreman to be employed for this work only, but with a view to employing him also later."

41. Ibid., Rutenberg's instructions concerning Jaffa consumers: "To be very strict with regard to payment"; for Jaffa Municipality: "Street Lighting: Try to obtain that such should be on the basis of payment for kWh used"; with respect to Tel Aviv: "With regards of payments from the Tel Aviv [Township], be very strict in observing terms."

42. TAHA file 2376-9, 1924–1925 documents relating to the proposed urban tramway.

43. The group was led by the commercial firm of Blattner & Co. in Alexandria, Egypt; see TAHA 2376-9, December 5, 1924, letter from Blattner to the Electric Company.

44. Meir Dizengoff, mayor of Tel Aviv, was at the same time a major partner in a large international shipping company. It was not always clear when he acted in his capacity as mayor and when in his capacity as a businessman in vigorously pursuing the tramway project. See, for example, his September 8, 1924, letter to the Electric Company, the letterhead of which bears the title of his commercial enterprise; compare this to his June 25, 1925, letter, carrying a similar appeal, under the letterhead of the Township of Tel Aviv (TAHA 2376-9). For more information on Dizengoff's shipping company, see the National Maritime Museum website, http://www.hma.org.il/Museum/Templates/showpage.asp?DBID=1&LNGID=1&TMID=84&FID=1753&PID=5153.

45. See Hughes on AEG investments in Germany (1983: 178). Drawing on statistics of electrical consumption in Berlin at the turn of the century, Hughes reports that of total consumption "light amounted to 28 percent; stationary motor power, 24 percent; and streetcar, or traction power, 48 percent" (ibid.: 190). Nye also reports that the supply of electricity to streetcars sometimes amounted to two-thirds of overall production, playing a role in the development of powerhouses and in the ability to expand the grid for other purposes as well (1990: 93).

46. Nye (1990), influenced in his analysis by Simmel and Goffman, refers to newspaper reading on streetcars as a new type of social barrier and a means to avoid a direct gaze. He also refers to racial confrontations and other less agreeable forms of interaction that accompanied the development of urban mass transit systems.

47. TAHA 2376-9, June 25, 1925, proposal submitted to the Jaffa Electric Company by Blattner & Co.

48. Ibid., June 25, 1925, letter from Dizengoff to the Electric Company.

49. BNA CO 814/11, January 27, 1925, document giving the Electric Company six months to accept an offer to build the Haifa funicular on terms that would be decided in consultation with the Department of Public Works. However, detailed plans and negotiations entered a final stage only during 1926 and 1927. See BNA CO 733/117.

50. On the rationalization of London transport in the 1920s and 1930s, see Forty (1986).

51. BNA CO 733/96, August 10, 1925, memorandum from Rutenberg to Norman Bentwich. Along similar lines, Rutenberg also demanded that the ordinance deal more strongly with "those who disturb, destroy, or incite to disturb or destroy the works erected for production, transmission and supply of electricity, and [order] penalties against unlawful diversion or use of electricity." On the decisive impact of the Electric Company on the legislative process, see Chapter 3.

52. The government's Electricity Concessions Ordinance, which went into effect in 1927, adopted the Electric Company's position in article 19; see "The Electricity Concessions Ordinance 1927," *Official Gazette of the Government of Palestine*, March 7, 1927.

53. The report specifically referred to the Jewish colonies of Rishon-le-Zion and Rehovot and to the Mikwe-Yisrael agricultural farm. These colonies consisted of private owners of land, typically cultivating oranges (see Chapter 5).

54. ECA 358/210, "Jaffa Electric Company Second Annual Report to the Board of Directors (1923/1924)."

55. ECA 0352-111, September 12, 1927, letter from Rutenberg's office to the Electric Company's Haifa office: "I am ordered by Mr. Rutenberg to inform you that he wants you to arrange classes in Arabic for the clerks and workers."

56. See BNA CO 733/93, June 2, 1925, letter from the Electric Company's London lawyers to the Colonial Office, announcing that their clients were ready to operate the Haifa powerhouse but "are unable to start the generation of electricity there before the definitive concession is signed."

57. See BNA CO 733/90, March 13, 1925, letter from High Commissioner Samuel to the Colonial Office. Also see BNA CO 814/3, February 17, 1925, Colonial Office memo regarding the need to justify operations in Haifa.

58. See a compilation of articles translated from Arabic in CZA 7416-5/S25: "The land has enough misfortunes, sufferings and obstacles in the way of its progress and prosperity"; "Every stone you throw at this scheme is an obstacle which impedes our advance to improvement and progress"; "Is obscurity better for us than light? Until when shall we be sunk in the obscurity of generations and dark hateful intolerance?"

59. ECA 0475, "Report of the Board of Directors to Shareholders of the Palestine Electric Company for 1925 (3rd Annual Meeting)."

60. Attempts to boycott the electrification project were short-lived. The Municipality of Jaffa contracted the Electric Company to supply streetlights five months after the first wired streetlights appeared in Tel Aviv. Contracts with other Arab towns followed later. Smith explains that the boycott failed because of British pressure, "divisions within the Arab ruling class," and the sheer desire for "modern street lighting" (1993: 122).

61. The Palestine Electric Company became the Israel Electric Corporation (IEC) in 1948. In 1967, when Israel occupied the Gaza Strip and the West Bank, the IEC became

the major provider of electricity to these areas. Cease-fire and partition lines, borders, checkpoints, fences, and separation walls notwithstanding, the IEC grid covers the whole territory of what used to be Mandatory Palestine. The Mavromatis Concession eventually became the East Jerusalem Electric Company, which supplied electricity to parts of the West Bank before 1967 and still operates today. At present there are three electricity distribution companies that operate in the West Bank: the Jerusalem District Electric Company (JDECO),which serves East Jerusalem, Jericho, Ramallah, and Bethlehem; the Northern Electricity Distribution Company (NEDCO), which operates in the northern West Bank; and the Southern Electric Company (SELCO), which serves the southern areas. The three companies purchase 95 percent of their needed electricity from the IEC and use a grid owned by it. The remaining 5 percent of electricity in the West Bank comes from Jordan. In Gaza, the IEC operates a power station that currently generates 70 megawatts, covering 30 percent of the 240-megawatt demand. Egypt supplies Gaza with 20 megawatts, and Israel supplies the remaining 150 megawatts. See http://export.gov/westbank/palestinianmarketbriefs/index.asp.

Two episodes may tell a bit about the current politics of electrification in Israel/Palestine. A *Jerusalem Post* news item of May 13, 2012, reported that Israel's Environment Protection Minister "called on the government to cut off the supply of electricity to the Gaza Strip in order to avoid electricity shortages it is feared could affect Israel this summer." Retrieved March 27, 2013, from http://www.jpost.com/Sci-Tech/Article.aspx?id=269715. Also see Weiss and Shamir (2011) for a socio-legal analysis of Israel's supply of fuel to Gaza. A second news item, from Ynetnews.com, October 6, 2010, reported that Jewish settlers in the city of Hebron (a Palestinian city on the West Bank) protested an IEC plan to construct a new high-tension line that would upgrade Hebron's electric supply. The official response was that the line would also feed Jewish settlements. Retrieved March 27, 2013, from http://www.ynetnews.com/articles/0,7340,L-3903479,00.html.

Chapter 3

1. TAHA file 39a-3.
2. BNA CAOG 14/109, September 21, 1921, Auja Concession, article 28.
3. On the connection of electrification to metrology, see note 38.
4. Sandra Sufian offers an excellent material history of malaria control in Palestine under British rule. She raises "fundamental questions about the relationship between technology, disease, and nationalism" (2007: 3) by showing how antimalaria practices were entangled with Zionism and the representation of the native Arab population and its land. Avitsur argues that the presence of malaria in the Auja River area shaped patterns of settlement in the vicinity until 1926–1927, when the Palestine Department of Health, in cooperation with the Tel Aviv Township, launched systematic antimalaria campaigns along the river (1957: 186).
5. BNA CAOG 14/109, article 30.

6. Ibid., article 12.

7. TAHA, file 39a-3.

8. Ibid., November 3, 1925, letter from Siemens to Tel Aviv Township recommending 200-watt lamps on uniquely designed armatures: "We hereby send you some samples for testing. Earlier we discussed flat covers but our experts think that this is not effective on five meter high poles because the light would be too strong to the eye. We now offer a different model, with a better dispersion of light from the reflectors. We also offer models that are not made of tin. They are more expensive but our experts think they would pay off in the long run. These models are less vulnerable to wind, they weigh less, and it is possible to install them in a way that would also throw light on the front of buildings. Sketches are attached and costs will be forthcoming" (translated from German).

9. Ibid., box 559, "Supply of Light 1924–1926." See, for example, the October 15, 1925, letter from the Police Department and the September 9, 1925, letter from merchants in the commercial center to Jaffa Municipality.

10. TAHA, file 6-48, January 19, 1926, Jaffa Electric Company notice.

11. TAHA, file 39a-3, November 22, 1925, meeting minutes.

12. The "light board" for October–December 1924, for example, specified that lamps be switched on between 5:35 p.m. and 7:30 p.m. on October 3–4; between 5:20 p.m. and 11 p.m. on October 7–15; and not switched on at all on October 25–31. See TAHA file 39a-3, "1924 Light Board."

13. For a rare anthropological study of "light and darkness in Palestine Folklore," see Canaan (1931). On the association of darkness (and even the mere absence of sunlight) with moral dangers, see Otter (2008: 67).

14. Lantern smashing is as old as the existence of streetlights. Schivelbusch suggests that this act of "vandalism" was often a form of rebellion against the order that streetlights embodied (1995: 98). In Jaffa, lantern smashing was associated with resistance to Jewish light. The Jaffa police force, under orders from the British governor, acted swiftly in such cases and also fined shop owners for failing to protect streetlights in their immediate vicinity.

15. TAHA, files 39a-3, 2-17, 3-40, September 14, 1924, and January 3, 1925, letters.

16. TAHA, file 39a-3, February 1, 1925, letter. The reference to new Jewish immigrants being particularly vulnerable at night surely invoked memories of the 1921 riots, when Arabs launched a deadly attack on the Jewish Immigration Center in Jaffa. On the association of "darkness" with the dangerous indigenous populations of colonial settings, see, for example, Malouf (1993) and the classic work of Joseph Conrad (1990 [1903]).

17. Otter developed a critique of panopticism as a dominant paradigm for studying the history of public spaces. Against the tyranny of the all-encompassing panoptic gaze that works at the service of surveillance and control, Otter argues for an oligoptic way of seeing, one that lacks a central vantage point and allows a symmetrical mutual gaze. His is

a liberal way of seeing things: local governments created infrastructures and shaped spaces that enhanced the abilities of citizens to see more clearly, encouraged "public self-observation" and private introspection, and allowed "sustained nocturnal perception" (2008: 63, 134). Others note how the spread of gas-lit lamps and later incandescent lamps eventually allowed the conquest of social time and the unprecedented displacement of the "night as frontier" (e.g., Melbin 1978).

18. TAHA, file 39a-3, October 1, 1925, letter from the Jaffa Electric Company to Tel Aviv Township.

19. BNA CO 733/91, March 1925, March 7, 1927, *Official Gazette of the Government of Palestine* on the Electricity Concessions Ordinance.

20. Ibid. See also CO 733/119, November 6, 1926, comments submitted to the Colonial Office by the Palestine Electric Company, and ECA 0357-184, October 28, 1926, correspondence between the Palestine Electric Company and the Department of Public Works.

21. BNA CO 733/96, August 11, 1925, memorandum from Norman Bentwich (chief legal adviser to the Government of Palestine) to the Chief Secretary of the Government.

22. Ibid., August 10, 1925, Rutenberg memorandum to Bentwich. One of the conclusions was that the 62-page draft ensured that "the number and amounts of penalties mentioned in the Ordinance would certainly largely increase the work of the law courts but not the supply of electricity."

23. BNA CO 733/96 9, August 10, 1925, letter from Rutenberg to Bentwich.

24. Ibid., August 11, 1925, Bentwich to the Chief Secretary of the Government; also, CO 733/119, August 8, 1925, dispatch no. 951 from Bentwich to Amery at the Colonial Office.

25. BNA CO 733/119, September 9, 1925, letter.

26. BNA CO 733/119, March 22, 1926, letter from Preece, Cardew & Rider to the Colonial Office. The firm had submitted an earlier three-page report, endorsed by Rutenberg, on March 3, 1926.

27. BNA CO 733/119, June 8, 1926, Colonial Office internal correspondence (Shuckburgh to Symes): "You may like to know that Rutenberg descended on me this morning. He was very anxious to track you down." Shuckburgh explained that Rutenberg told him that he wanted to advance the Ordinance and that he fully trusted and submitted to the opinion of Preece, Cardew & Rider.

28. BNA CO 733/116, August 8, 1926, telegram further explaining that the government also borrowed from the India Electricity Act of 1910.

29. TAHA, File 39a-3, September 6, 1925, letter warning that a yet-unconnected property had a pharmacy on its premises and that it was dangerous to keep oil lamps in close proximity to medicinal chemicals. A file full of letters of this sort confirms that after years of their use, oil lamps, suddenly became a serious hazard. Also see TAHA file 3-40, September 28, 1924, letter from residents of Tel Nordau to Tel Aviv Township.

30. The petition was submitted by residents of the affluent neighborhood of Tel Nordau. On the history of this precinct, see Gophna (2008).

31. TAHA file 3-40, September 28, 1924, letter from residents of Tel Nordau to Tel Aviv Township.

32. Nye (1990), Otter (2008), Schivelbusch (1988). Walter Benjamin (1999) collected impressions and reflections about public light as part of his larger effort to outline "the materialist philosophy of the 19th century" (Jenkins and Munslow 2004: 135).

33. ECA 0475, "Palestine Electric Corporation Report of Directors for the Year Ended December 31, 1933."

34. Still, the overall provision of "light" for both public and private consumption (including water supply) was considerably lower than the provision of electric power to industry and irrigation. See Chapter 5.

35. ECA 0358-210, "Jaffa Electric Company, First Report of Directors for 1923–1924" and "Jaffa Electric Company, Second Report of Directors for 1924–1925."

36. ECA 0475, "Palestine Electric Corporation, Report of Directors for the Year Ended December 31, 1933." Also see TAHA, file 39a-3, October 1, 1925, Electric Company report to Tel Aviv Township. Data collected by Faris (1936) show that the number of individual consumers in Tel Aviv grew from 5,324 in 1927 to 10,401 in 1932. Over the same period, the number of consumers grew from 1,906 to 4,264 in Haifa and from 227 to 448 in Tiberias. The data also show phenomenal growth in profits. For more detailed data, see CZA 185/46, December 31, 1925, "Jaffa Electric Company, Report of Directors for the Year 1923/1924."

37. See Otter (2008) on the standardization of light units and the important role played by prepayment gas meters for the spread of gas-lit networks.

38. Rates were based on three principal categories: residential/commercial, street lighting and water supply, and irrigation and industry. The original rates were fixed at 5 Egyptian piastres (PT) for private consumers, 2.5 PT for street lighting and water supply, and 2 PT for irrigation and industry (for currencies, see note 8 in Chapter 1).

39. The importance of aligning metrology with other actor networks is convincingly illustrated by Shaffer (1997), who traces the role of accepted standards in facilitating industrial development, global trade, and the creation of telegraph systems.

40. "Boundary objects" are useful instruments in the toolbox students of actor networks employ—namely, in the way actor networks are assembled, stabilized, and maintained. Some users of the term distinguish several types of boundary objects (e.g., Star and Griesemer 1989; Fleischmann 2006), while others deploy it in a generic sense (e.g., Carroll 2012).

41. ECA 0357-184, *Jaffa District Electric Works, General Conditions for the Connection and Supply of Electrical Energy*, chap. 8.

42. BNA CAOG 14/109, Auja Concession, article 12.

43. BNA FO 141/634/4, November 18, 1926.

44. BNA FO 141/634/4. The companies were Compagne Centrale D'eclairage et de Chauffage par le Gaz (Alexandria); Helouan Local Commission Electrical Services (Helouan); Cairo Electric Railways and Heliopolis Oasis Company (Cairo); Electricity and Ice Supply Company (Ismailia); and Electricity and Ice Supply Company (Suez and Port Tewfic). The questionnaire was sent to the companies by the secretary of the High Commissioner for Egypt on November 24, 1926.

45. BNA FO 141/634/4, November 24, 1926.

46. Ibid., November 30, 1926.

47. "Electricity Concessions Ordinance of 1927," *Official Gazette of the Government of Palestine*, March 7, 1927.

48. ECA 0357-184, *Jaffa District Electric Works, General Conditions for the Connection and Supply of Electrical Energy*, chap. 9.

49. Ibid., chaps. 6 and 11.

50. Ibid., chap. 12.

51. ECA 2376-17, June 23, 1926, summary of union meetings. Also see PLA files iv-250-72-1-2572 and iv-250-72-1-2100.

52. See article 25 of the Jordan Concession, incorporated in "Electricity Concessions Ordinance," *Official Gazette of the Government of Palestine*, March 7, 1927. Relations had begun with the signing of an application form. Applications included a one-year commitment to remain wired to the main distribution system and various obligations to safeguard electric components and to submit to the technical authority of the company. The signing of the application form and the subsequent installment of a current meter was designed to initiate a lifelong contractual relationship.

53. See Gomart and Hennion, who argue for an "event-network-theory" (1999: 221, 227), focusing on modalities of passionate attachment (music lovers and drug users) as a way to discuss "the mix between active and passive" and the irreducibility of the actor into a subject or an object (1999: 222). In such terms, the availability of wired electrical current was an event that bred an "attachment" and an "opportunity to be seized." For another, related sociological analysis of the subject, see Moser and Law (1999).

54. Of course, this egalitarian property must be seriously qualified. Other chapters of this book show that the direction and velocity of the grid depended on and in turn dictated considerable asymmetry among geographical regions and specific localities. The grid displayed a rationale of expansion that was far from egalitarian, answering to variable levels of demand and variable levels of willingness and capacity to invest in its expansion. It is only here, discussing the way grids produced and consisted of public and private regions, that the term "egalitarian" is applied in a restrictive sense.

55. The distinction between objects' constitutive and symbolic functions is subtle yet critical. Durkheim's observation may be useful here: "The emblem is not merely a convenient process for clarifying the sentiment the society has of itself: it also serves to create this sentiment; it is one of its constituent elements" (1947 [1915]: 230–231). A totem, a

legal system, a regal emblem, a national football team, or an electric lamp not only expresses this or that social group but actually allows it to exist.

56. See Hughes: light "was not for the poor" in nineteenth-century Europe (1983: 73).

57. ECA 0358-210, "Jaffa Electric Company, Second Report of Directors, 1924–1925."

Chapter 4

1. BNA T161/899. Apart from the rail lines, the British Army also seized 26 German locomotives. Six were lent to the Damascus–Hama Railway Company, five were in use in Palestine, and fifteen operated on the Hedjaz line. Also see BNA CO 733/13.

2. Ibid., file containing plans to help King Ibn-Saud reconstruct the line to Medina. Also see BNA CO 733/80 and CO 733/81.

3. BNA CO 733/13, July 1, 1921, memorandum from the War Office to the Undersecretary of State, expressing the need to determine the future control of the Hedjaz line.

4. BNA CO 733/35, July 1922 memorandum to the Colonial Office summarizing the position of Colonel Holmes, general manager of the Palestine Railway: "We must obtain control of the Semakh-Deraa section. At present we have to pay the French 30 English Pounds for each train running this line and this is 'blackmail payment.' Palestine Railway runs the line but Trans-Jordan meets the expenses and gets the revenue. Also, we need to open the line all the way down to Medina under Palestine Railway but we need money for that and the French may not consent." The British Foreign Office approached the Colonial Office in 1924, wanting to know whether the British government or private British firms could become owners of the line, because "it would give an opportunity of opening up some trade for Britain." The Foreign Office agreed that the venture would be highly important for relations with King Ibn-Saud and would improve the standing of the colonial powers with respect to the "whole pilgrimage issue." A latter-day document explained that the effort failed because Ibn-Saud refused to engage in technical matters prior to a final decision on the ownership and "unity" of the line. A final abortive attempt to conclude principled issues with respect to the Hedjaz line was made in late 1935, when at a conference at Haifa it was agreed to restore the line to Medina and, again, to establish a committee composed of three French representatives for Syria, three British representatives for Palestine and Trans-Jordan, and four Saudi representatives. The wedding of the Hedjaz to an electric grid was never on the agenda.

5. BNA CO 733/3. The Ottoman concession to build and operate the railway line between Jaffa and Jerusalem had been granted to the French company on October 29, 1888. It was to be in effect for seventy-one years, until 1959.

6. Ibid., JJRW arbitration file. On May 4, 1921, the High Commissioner informed the Colonial Office that, with the failure of agreement on the sums to be paid, he had written the French General Consul that the claim had been passed to London. The Foreign Office was also involved in the negotiations that followed.

7. BNA CO 814/3, September 22, 1922, "Minutes of the Executive Council of the Government of Palestine, Discussion of the JJRW arbitration." The authorization of the Colonial Office came after "experts convened" on the matter and concluded that it was better not to await the award of the tribunal. The Executive Council resolved to ensure that the JJRW would remain the property of the government. It also resolved to increase the sum of a loan it had requested from the Treasury in London, specifying that it had only £300,000 at its disposal toward that end.

8. BNA CO 733/3, May 15, 1921, memorandum from Rutenberg to Samuel.

9. Ibid., June 3, 1921, dispatch no. 142 from Samuel to the Colonial Office. Enclosed was another memorandum from Rutenberg, repeating the claim that the electrification of the JJRW "is a necessary condition of the completion of the hydroelectric scheme."

10. Ibid., June 7, 1921, dispatch no.144 from the Department of Commerce and Industry to the Colonial Office.

11. Ibid., July 22, 1921, report submitted by Preece, Cardew & Rider to the Colonial Office on the feasibility of the Jordan Project and the electrification of railways (hereafter Preece Report).

12. Reguer (1995) also observed that the Jordan Project depended on the massive sale of electricity to the biggest buyer and that the only enterprise that could draw so much electricity was the railways. Also see Shaltiel (1990: 90).

13. On the involvement of Snell, formerly a senior partner of Preece & Associates (which became Preece, Cardew & Rider after Snell's death), see Chapter 1.

14. BNA CO 733/3, May 15, 1921, memorandum from Rutenberg to the High Commissioner.

15. Ibid., 2–3.

16. Ibid., 4.

17. Ibid.

18. Ibid., 5.

19. Ibid., June 3, 1921, dispatch no. 142 from High Commissioner Samuel to Winston Churchill at the Colonial Office.

20. Ibid., June 7, 1921, unnumbered dispatch from the Department of Commerce in Jerusalem to the Colonial Office.

21. Ibid., July 22, 1921, Preece Report.

22. BNA CO 733/7, November 11, 1921, letter from the Colonial Office to the High Commissioner of Palestine, informing him that the consulting engineers had found that the technical plan was sound but not financially advantageous.

23. See the opinion of Mr. Caulson, cited by Reguer (1995: 700). Herbert Samuel was also aware of this technological competition and therefore emphasized in his messages that the experts must weigh the benefits of electrification against the "alternative of a conversion of the existing locomotives for the consumption of oil fuel." See BNA CO 733/3, June 3, 1921, dispatch no. 142 from High Commissioner Samuel to Winston Churchill at the Colonial Office.

24. Reguer (1995: 700), quoting Mr. Vernon at the Colonial Office.

25. BNA CO 733/3 and Reguer (1995: 700). Rutenberg tried to commit the government to take at least 4 million kilowatt-hours annually for the JJRW.

26. Reguer (1995: 700). These critical comments referred to the fact that Rutenberg took the initiative of soliciting tenders from British and German firms for those parts of the electrification project that he expected the government to undertake.

27. BNA CAOG 14/109, August 10, 1921, letter from Rutenberg to Mr. Vernon at the Colonial Office.

28. Only after further negotiations and five more years would the Jordan Concession for the hydroelectric electrification of Palestine be formally signed. BNA CAOG 14/109, September 22, 1921, letter (46648/21) from the Colonial Office gave instructions not to approve the contract "by which the Palestine Government binds itself to purchase a fixed annual amount of electrical energy."

29. BNA CAOG 14/109, December 23, 1921, letter from Preece, Cardew & Rider to the chief engineer of the Crown Agents for the Colonies.

30. Ibid., March 1, 1922.

31. BNA CO 814/1, Department of Public Works report no. 1.4.24-31.3.25.

32. BNA CO 733/29, November 11, 1921, "Report of Consulting Electrical Engineers"; February 10, 1922, memorandum; and March 23–29, 1922, letters. I thank Fredrik Meiton for pointing out these files to me. Also see BNA CAOG 14/109.

33. I could not find archival evidence of the early refusal and then the circumstances that led the Air Ministry to change its position on that matter. I rely here on the undocumented claims of Yahav (2009: 82).

34. ECA 352/111, August 16, 1924, "Instructions for the Erection of Sarafand High-Tension Line." Rutenberg issued minute instructions and schedules for purchasing land, settling disputes over passage through private land, and building transformer stations and the high-tension line, demanding that the project be concluded by the end of 1924.

35. BNA TS 28/124.

36. Ibid., undated, "Jaffa Electrical Company Agreement for Supply of Electricity to Military Camps at Sarafand." This file contains drafts and memos from 1924–1925 regarding an agreement between the company and the army.

37. Ultimately, the agreement between the parties also permitted a special schedule of rates on the basis of a diminishing sliding scale. The strictly commercial aspect of the agreement was buried beneath technical and legal considerations, but it should not be ignored. In the 1890s, the British in India established a Royal Commission of Inquiry to look into the cultivation and consumption of hemp. Its final report recommended the legalization of the trade, offering a fascinating actor network that brought together government statistics, mental illness experts, public surveys, economists, political philosophies, and folklore. An Indian member of the commission wrote a minority opinion, arguing that this impressive assemblage was nothing but an attempt to legitimize the taxing appetite of the Crown. See Shamir and Hacker (2001).

38. The Auja Concession specifically referred to the relatively large Jewish settlement of Petah-Tikva, about 10 kilometers east of Jaffa.

39. The negotiations between the Electric Company and the organized local council of Rishon Le-Zion represent a strong example of the strictly commercial considerations that drove the grid. Negotiations began in early 1925, but concluded only in early 1927. The company demanded a substantial sum for branching off wires from the high-tension line (8,000 Egyptian pounds) and refused to settle in spite of pleadings and pressures. The contract was eventually concluded through a loan from the Anglo-Palestine Bank guaranteed by the government. The Council of Rishon Le-Zion also imposed a special tax on each resident to cover its interest payments. See RLZ, A 1-31 and P 2-1/1, protocols for assignment of poles and lamps, and a March 27, 1927, invitation to the ceremony commemorating the first day of electricity. A contract with Ramleh for the provision of streetlights was signed only later, toward the end of the 1920s. See Chapter 2 for more details on these particular locations.

40. ECA 358-210, December 31, 1925, Electric Company's annual directors' report, 2.

41. BNA CO 733/3, June 7, 1921, dispatch no.144 from the Department of Commerce and Industry to the Colonial Office.

42. BNA CO 733/129/13, August 5, 1926, letter from the Trade Facilities Advisory Committee to the Colonial Office. The expert opinion to which the letter refers was the Merz and McLellan report of 1923.

43. Ibid., July 13–26, 1926, Colonial Office internal memoranda summarizing Rutenberg's proposals. Also see a July 16, 1926, letter from Rutenberg to Sir John Shuckburgh at the Colonial Office, and a July 23, 1926, letter from the Palestine Electric Company to the Secretary of State for the Colonies,.

44. Ibid., July 27, 1926, confidential letter from Secretary of State Wilson to Sir William Plender, chairman of the Trade Facilities Advisory Committee. The letter emphasized "a great deal of importance" and the "utmost importance" of the loan authorization. Also see ibid., August 13, 1926, draft letter from the Colonial Office to the committee stating "how disappointed [the Secretary of State] was to learn of your Committee's unfavorable decision."

45. Ibid., August 6, 1926, letter from Alfred Mond to William Plender, chairman of the Trade Facilities Advisory Committee, expressing dismay at the latter's pessimism.

46. Ibid., August 13, 1926, draft letter.

47. Ibid., August 17–18, September 23, October 18, and November 5, 1926, correspondence. See BNA CO 733/137/6 for later legal complications owing to the status of Palestine and Trans-Jordan under the Mandate. The Jaffa Electric Company was incorporated into the Palestine Electric Company in 1927.

48. The Jewish immigration wave of the 1930s overwhelmed the earlier one, thereby obscuring some of the crucial processes that took place in the 1920s discussed in this chapter.

49. BNA CO 733/128/2, 1925 (otherwise undated and unsigned), internal government summary.

50. ECA 358-210, December 31, 1925, Electric Company's annual directors' report.

51. CZA 7416-71/S25, ca. August 1925 Jaffa Electric Company "Confidential Report." The table showed demand in kilowatts for "lighting for residences," "lighting of streets," "industrial power," and "water supply power." It measured the percentage increase in the "last nine months" (October 1924 through July 1925) and in the "last 21 months" (October 1923 through July 1925).

52. The kilovolt-ampere measures the output of a transformer before power is actually consumed. Converted to kilowatts, production capacity grew from 160 kilowatt-hours (200 kilovolt-amperes) to 1,000 kilowatt-hours (1,250 kilovolt-amperes).

53. See for example, TAHA a-39-3, December 30, 1924, letter from Tel Aviv Municipality to the Electric Company asking for extensions for 4 kilometers of new wires; October 26, 1925, letter from the company to Tel Aviv Township reporting on 6,736 meters of new township-ordered extensions and 2,764 meters of new nonmunicipal orders; and December 31, 1925, letter from the company to Tel Aviv Township reporting on further new extensions.

54. TAHA a-39-3, January 18, 1926, letter and appendices from the Electric Company to Tel Aviv Township.

55. ECA 0358-210, December 31, 1925, Electric Company's annual directors' report.

Chapter 5

1. CZA 7416-20/S25, June 19, 1924, letter from Kisch to Weizmann.

2. CZA 7416-30/S25, June 26, 1924, letter from the Electric Company to the Palestine Executive of the Zionist Organization in Jerusalem.

3. CZA 7416-45/S25, July 23, 1924, letter from the Electric Company to the Zionist Executive.

4. BNA CO 814/3, June 30, 1924, minutes of Palestine Executive Council meeting.

5. Ibid., July 30, 1925, minutes of Palestine Executive Council meeting; August 21, 1925, final recommendation.

6. BNA CO 733/3, June 7, 1921, dispatch no.144 from the Department of Commerce and Industry to the Colonial Office.

7. See Chapter 3 on the role of current meters in distinguishing the public and the private and on stratifying the private.

8. CZA 7416-71/S25. The table showed demand in kilowatt-hours for "lighting for residences," "lighting of streets," "industrial power," and "water supply power." It measured the percentage of increase in the "last nine months" (October 1924 through July 1925) and in the "last 21 months" (October 1923 through July 1925).

9. Ibid. The exact numbers were 8,352 kilowatt-hours in October 1923, 29,940 in October 1924, and 68,464 in July 1925. The growth was remarkable in the other categories

as well: between October 1923 and July 1925, demand for residential lighting grew by 260 percent, demand for streetlights grew by 160 percent, and demand for water supply power grew by 380 percent.

10. ECA 0475, May 21, 1925, Palestine Electric Corporation annual directors' report. Based on its early figures for 1924–1925, the company stated that "the reports of the [company] are showing good progress. The quantity of energy sold has increased from January 1924 to April 1925 by 456 percent."

11. Ibid., December 31, 1927, Palestine Electric Corporation annual directors' report and balance sheet: "Agreements have been entered into with the Jaffa Electric Company Ltd for the incorporation of the Auja concession in the Jordan concession and a granting of a license to the Jaffa Electric Company Ltd for generation, distribution, and sale of electric energy in the Auja Concession area."

12. Ibid., December 31, 1935, Palestine Electric Corporation annual directors' report and balance sheet.

13. Ibid., December 31, 1933, Palestine Electric Corporation annual directors' report and balance sheet.

14. The exact figures provided by Metzer indicate growth from a ratio of about 0.5 in the early 1930s to 2.0 in 1939 for Palestine as a whole, and from 0.4 in 1921–1922 to 2.5 in 1930 and then to 5.1 in 1937 for the Jewish sector.

15. Some of the industries in Haifa were significantly bigger in size and attracted £665,000 in capital investments while the share of Tel Aviv industries was £454,000. The difference between the number of new industries and the volume of investment between the two towns (lying 100 kilometers apart) had to do with the specific prospects of Haifa, which—according to British design and plans—was to become a major terminal for oil arriving by pipelines from Mosul (in present-day Iraq). The projection for Haifa included a major port, crude oil refineries, and related industries (e.g., petrochemicals). A surrounding ring of other relatively heavy industries had also begun to concentrate in Haifa, with flour mills, cement works, and salt and cooking oil factories already in production. See BNA DO 118/66 for the concession agreements between the Government of Palestine and oil companies in the 1930s and 1940s: "Convention with the Iraq Petroleum Company and the Construction of Refineries in Haifa," "Convention with the Anglo-Persian Company," and the 1940 agreement with Consolidated Refineries.

16. A second and smaller powerhouse and a distribution system had been built at the same time in the town of Tiberias. See ECA 0475, May 21, 1925. The Palestine Electric Corporation's annual directors' report specified that the supply of electricity began on September 1, 1925, at Haifa and on November 5, 1926, at Tiberias. Also see BNA CO 814/3, Department of Public Works annual report for 1925, finding both powerhouses "to be run and kept in first class condition."

17. ECA 358-210, December 31, 1925, Jaffa Electric Company annual directors' report, 2.

18. CO 733/129/13, August 6, 1926.

19. Callon writes about the "performativity of economics" in framing the calculable domain of the economy and in creating the institutions that engage in rational politics of calculation. For further developments of this approach, see MacKenzie, Muniesa, and Siu (2008).

20. See Meyer and Rowan (1977) for the actor networks that generate rationalized institutional contexts.

21. Julie Trottier follows a definition of hydro-politics as "the investigation of water conflicts in order to uncover the tensions among competing interests, as well as the types of political, imaginary and symbolic relations which the issue of water mobilizes" (1999: 3); also see Waterbury (1979). From the perspective of ANT, Alatout inquires "how exactly water is enrolled in politics and with what effects" (2009: 371).

22. Article 7 of the Auja Concession specifically referred to the Jewish settlement of Petah-Tikva, which was located roughly 10 kilometers east of Jaffa, not far from the source of the Auja (Yarkon) River. Established in 1878 and based on private farming, Petah-Tikva was for many years the biggest center of Jewish citrus farming in the number and size of its planted orchards.

23. ECA 358-210, December 31, 1933, Palestine Electric Corporation's annual directors' report and balance sheet.

24. See Chapter 1.

25. The circumstances surrounding the Mavromatis Concession are discussed in detail in the Introduction to this book.

26. ECA 0475, June 29, 1926, Palestine Electric Company annual directors' report and statement of accounts, referring to the decision of the Permanent Court of International Justice of March 26, 1925. The report claimed that "the experts nominated by the two parties instead of adapting the old concession to the present economic conditions elaborated an entirely new concession, going, in our opinion, considerably beyond the decision of the Court thus affecting the interests of the Palestine Electric Corporation." The board was told that the company had lodged a complaint in this matter and intended to defend its interests.

27. BNA CO 733/92, May 18, 1925, dispatch no. 509 from Amery in the Colonial Office to Herbert Samuel; June 8, 1925, follow-up letter to the Jaffa Electric Company. This demand was based on the original commitment to supply electricity to Petah-Tikva, as yet unfulfilled in mid-1925. There were also earlier signs that the government tried to find a way to regain control over the water rights in the Auja. See February 17, 1925, minutes of the Palestine Executive Council meeting, in which there was no objection to extending the original obligation for another six months, with the understanding that if Rutenberg were to be eventually relieved of this obligation he "should not be permitted to utilize the Auja River to generate power for irrigation purposes."

28. BNA CO 733/96, August 12, 1925, memorandum from High Commissioner Samuel to the Colonial Office, informing it that a hydroelectric station had become "totally unnecessary."

29. Ibid.

30. BNA CO 733/124, April 16, 1926, letter from the Jaffa Electric Company to the Colonial Office.

31. BNA CO 733/116, August 18, 1926, Department of Public Works annual report submitted to the Chief Secretary of the Government, marked "confidential/urgent." The report offered calculations on the basis of the Electric Company's claim that the irrigation of orange groves required 2,000 cubic meters of water a day for 10,000 dunam (2,500 acres). Acting on this figure and relying on other available studies, the report found that the area that could be irrigated around the Auja was about 100,000 dunam. Reducing this figure by 20 percent for lands unsuitable for irrigation, it established that the irrigable area would in fact be 8,000 dunam. If all of this area were to be planted with orange groves, the report suggested, it would require 160,000 cubic meters of water a day, according to the Electric Company's own calculation. Other studies showed that the Auja could provide 735,000 cubic meters a day during its lowest season of flow, an amount still four times higher than what would have been required for irrigation.

32. BNA CO 733/134/6, January 19, 1927, letter from the Law Officers Department at 10 Downing Street to the Colonial Office.

33. BNA CO 733/124, August 18, 1926, letter from the Law Officers Department at 10 Downing Street to the Colonial Office.

34. Rutenberg kept negotiating the fine details of the deal throughout late 1926 and 1927. See BNA CO 733/134/6, December 28, 1926, letter from the Electric Company to the Chief Secretary in Jerusalem: the company agreed to allow the supply of 20,000 cubic meters daily from the waters of the Auja to Jerusalem, with the understanding that the government would not extend the Mavromatis Concession further than what it was obliged to and that the scheme of water supply would include the creation of "large reservoirs to render possible pumping of water during night time so as not to prejudice the use of water for irrigation." It is not clear whether these demands were met by the government. A draft agreement formally releasing Rutenberg from the obligation to build a hydroelectric powerhouse was still in circulation as late as October 1927.

35. BNA CO 733/118, December 2, 1926, letter from the Colonial Office to Euripides Mavromatis authorizing his plan for water supply in Jerusalem. An earlier telegram of November 26, 1926, from the High Commissioner in Jerusalem to the Colonial Office stated that Rutenberg had been informed of the decision.

36. BNA CO 733/96, July 24, 1925, letter from Rutenberg to Herbert Samuel.

37. ECA 358/210, December 31, 1925, Jaffa Electric Company annual directors' report.

38. It seems that the government's approval of the plan was delayed because of an unexpected interruption by science. The Hope-Simpson Commission found that the government's technical experts had had no time to examine and approve the plan "in consequence of the occupation of those advisers on the locust campaign." It also noted

that "the original concession was given in the year 1922. Eight years have passed and so far nothing has been done. It is desirable that the scheme put forward by the concessionaires last year should be examined and reported upon as soon as that is possible. The sooner irrigation is available for the plain the better" (1930: 83). See "Report on Immigration, Land Settlement and Development By Sir John Hope Simpson, C.I.E., Presented by the Secretary of State for the Colonies to Parliament by Command of His Majesty, October, 1930" (the Hope-Simpson Report). The full text of the report is available at http://www.zionism-israel.com/Palestine_Hope_Simpson_Report.htm.

39. Hope-Simpson Report (1930).

40. See the Hope-Simpson Report (1930) on the conflicting findings concerning available irrigable land offered by experts employed by the Jewish Agency and by Dr. Strahorn, who advised the government. See El-Eini (2006) for a detailed account of the legislative efforts of the Palestine government, the history of water and agriculture legislation in the 1930s and the 1940s, and a discussion of the ability of Zionist institutions to prevent or shape and amend the emergent water regime.

41. Citrus products amounted to 43 percent of total exports in 1927 and 84 percent of exports in 1935 (Karlinsky 2000: 328; Rokach 1970).

42. The biggest and most established colony was Petah-Tikva (founded in 1878), which largely depended on citrus farming and was considered by both the British government and the Electric Company as a potentially important consumer of electricity for irrigation. Other notable relatively "old" Jewish colonies that depended on privately owned orchards were Rishon Le-Zion (founded in 1882), Nes-Ziona (founded in 1883), Rehovot (founded in 1890), and Kfar-Saba (founded in 1903).

43. I draw this conclusion from considering a map showing electric extensions from the high-tension line to Sarafand to neighboring settlements and wells. Dated November 11, 1928, this map marked the names of individual well owners, revealing their ethno-national identity. It showed extensive wiring to Jewish-owned wells as well as some scattered connections to Arab-owned ones. The grid did not cater exclusively to Jewish owners, and this fact only strengthens my overall argument: the grid did not follow any ideological or political ethno-national bias but simply obeyed the logic of attachment to those who were able to invest capital in electric pumping. See ISA, Map DEST 3408.

44. When the Hope-Simpson Commission began its inquiries in late 1929, it found sound "agricultural work of a scientific character" being undertaken by Zionist institutions and Jewish scientists. The report singled out the Jewish Agency and the Palestine Jewish Colonization Association (PICA), specifically detailing "research work of the highest value" undertaken by the Agricultural Experimental Station of Tel Aviv. With a staff of 40 scientists, the experimental station had three laboratory divisions (plant pathology, entomology, agricultural chemistry) and divisions such as agronomy, plant breeding, and horticultural breeding (1930: 76). The report also gave special thanks to Dr. Jizhaq Wilkansky (Itzchak Volkani-Elazari) and his work, *The Fellah's Farm* (1930) (originally

published in Hebrew in 1925). Finally, it mentioned the scientific contributions of chemists and analysts at the Hebrew University.

45. This would change a few years later, when considerable waters had begun to be electrically pumped from both the Auja and the Jordan. The Palestine Electric Company created a separate irrigation company for the Auja in 1929 and later it also created several pumping stations on the Jordan River.

References

Achenbach, Joel (July 2010) "The 21st Century Grid: Can We Fix the Infrastructure That Powers Our Lives?" *National Geographic*. Retrieved March 3, 2012, from http://ngm.nationalgeographic.com/2010/07/power-grid/achenbach-text.

Alatout, Samer (2009) "Bringing Abundance into Environmental Politics: Constructing a Zionist Network of Water Abundance, Immigration, and Colonization." *Social Studies of Science* 39, 363–394.

Ascherson, Neal (1963) *The King Incorporated: Leopold the Second and the Congo*. London: George Allen & Unwin.

Atwood, Margaret (2000) *The Blind Assassin*. New York: Anchor Books.

Avitsur, Shmuel (1957) *The Yarkon: The River and Its Environment*. Tel Aviv: Hakibbutz Hameuchad [in Hebrew].

Baker, C. Edward (1980) *Preece and Those Who Followed: Consulting Engineers in the Twentieth Century*. Brighton: Reprographic Centre.

Barnes, Douglas F. (1988) *Electric Power for Rural Growth: How Electricity Affects Rural Life in Developing Countries*. Boulder, CO: Westview Press.

Barton, A. Gregory, and Bennett, M. Bret (2010) "Forestry as Foreign Policy: Anglo-Siamese Relations and the Origins of Britain's Informal Empire in the Teak Forests of Northern Siam, 1883–1925." *Itinerario* 34, 65–86.

Beaulieu, Anne (2010) "From Co-Location to Co-Presence: Shifts in the Use of Ethnography for the Study of Knowledge." *Social Studies of Science* 40(3), 453–470.

Beilin, Yossi (1987) *The Origins of Jewish Industry*. Jerusalem: Keter [in Hebrew].

Ben-Eliezer, Uri (1995) *The Emergence of Israeli Militarism 1936–1956*. Tel Aviv: Zmora-Bitan [in Hebrew].

Benjamin, Walter (1999) *The Arcades Project*. ed. Rolf Tiedemann. Cambridge, MA: Harvard University Press.

Ben Porat, Amir (1999) *The Bourgoise: The History of the Israeli Bourgeoises*. Jerusalem: Magnes Press [in Hebrew].

Bernstein, Deborah (2000) *Constructing Boundaries—Jewish and Arab Workers in Mandatory Palestine*. Albany: State University of New York Press.

Biger, Gideon (1983) *A Crown Colony or a National Home: The Impact of British Rule on Palestine 1917–1930, a Historical-Geographical Account*. Jerusalem: Ben Zvi [in Hebrew].

Borchard, Edwin M. (1925) The Mavromatis Concessions Cases. *American Journal of International Law* 19(4), 728–738.

Busch, Lawrence, and Juska, Arunas (1997) "Beyond Political Economy: Actor Networks and the Globalization of Agriculture." *Review of International Political Economy* 4(4), 688–708.

Buzelin, Helene (2005) "Unexpected Allies: How Latour's Network Theory Could Complement Bourdieusian Analyses in Translation Studies." *The Translator* 11(2), 193–218.

Callon, Michel (1999) "Actor Network Theory—The Market Test." In: J. Law and J. Hassard (eds.) *Actor Network Theory and After*. Oxford: Blackwell/Sociological Review, 181–195.

——— (2005) "Why Virtualism Paves the Way to Political Impotence: A Reply to Daniel Miller's Critique of the Laws of the Markets." *Economic Sociology European Electronic Newsletter* 6(2), 3–20.

Callon, Michel, Lascoumes, Pierre, and Barthe, Yannick (2009) *Acting in an Uncertain World: An Essay on Technical Democracy*. Cambridge, MA: MIT Press.

Canaan, Tawfiq (1931) "Light and Darkness in Palestine Folklore." *Journal of the Palestine Oriental Society* 16, 15–36.

Carroll, Patrick (2012) "Water and Technoscientific State Formation in California." *Social Studies of Science* 42(2), 313–321.

Chambers, Robert (1994) "The Origins and Practice of Participatory Rural Appraisal." *World Development* 22(7), 953–969.

Chelouche, Yosef (2005 [1931]) *Reminiscences of My Life 1870–1930*. Tel Aviv: Babel [in Hebrew].

Chikowero, Moses (2007) "Subalternating Currents: Electrification and Power Politics in Bulawayo, Colonial Zimbabwe, 1894–1939." *Journal of Southern African Studies* 33(2), 287–306.

Conrad, Joseph (1990 [1903]) *The Heart of Darkness*. Edinburgh: Dover Thrift Editions.

Coquery-Vidrovitch, Catherine (2003) "Electricity Networks in Africa: A Comparative Study, or How to Write Social History from Economic Sources." In: T. Falola and C. Jennings (eds.) *Sources and Methods in African History: Spoken, Written, Unearthed*. Rochester, NY: University of Rochester Press, 346–360.

Cotterell, Paul (1989) "A Tramway Project in Jaffa." *HaRakevet* 6, 11–12 [in Hebrew].

Dothan, Shmuel (1979) *Partition of Eretz-Israel in the Mandatory Period: The Jewish Controversy*. Jerusalem: Ben-Zvi [in Hebrew].

Durkheim, Emile (1947 [1915]) *The Elementary Forms of the Religious Life: A Study in Religious Sociology*. Glencoe, IL: Free Press.

Elazari-Volkani, Isaac (1925) "The Transition from Primitive to Modern Agriculture in Palestine." *Journal of the Palestine Economic Society* 1.

El-Eini, Roza (1996) "The Implementation of British Agricultural Policy in Palestine in the 1930s." *Middle Eastern Studies* 32(4), 211–250.

——— (1997) "Government Fiscal Policy in Mandatory Palestine." *Middle Eastern Studies* 33(3), 570–596.

——— (2006) *Mandated Landscape: British Imperial Rule in Palestine, 1929–1948*. London: Routledge.

Ewans, Martin (2002) *European Atrocity, African Catastrophe: Leopold II, the Congo Free State and Its Aftermath*. London: Routledge Curzon.

Faris, A. Basim (1936) *Electric Power in Syria and Palestine*. Beirut: American University of Beirut Press.

Fellman, Aharon L. (1891) *Ma'ayan Ganim* [*Spring of Gardens*]. Jerusalem: R.A.M. Luncz Printery [in Hebrew].

Fleischmann, Kenneth R. (2006) "Boundary Objects with Agency: A Method for Studying the Design-Use Interface." *Information Society* 22, 77–87.

Forty, Adrian (1986) *Objects of Desire: Design and Society since 1750*. London: Thames & Hudson.

Furlong, Kathryn (2010) "Small Technologies, Big Change: Rethinking Infrastructure through STS and Geography." *Progress in Human Geography* 35(4), 1–23.

Giladi, Dan (1971) "The Economic Crisis during the Fourth Aliyah (1926–1927)." *HaZionut* [*Zionism*] 2, 119–147 [in Hebrew].

——— (1973) *The Yishuv during the Fourth Aliyaa 1924–1929: A Political and Economic Study*. Tel Aviv: Am Oved [in Hebrew].

Gilbert, Martin (2007) *Churchill and the Jews: A Lifelong Friendship*. London: Simon & Schuster.

Gomart, Emilie, and Hennion, Antoin (1999) "A Sociology of Attachment: Music Amateurs, Drug Users." In: J. Law and J. Hassard (eds.) *Actor Network Theory and After*. Oxford: Blackwell, 220–247.

Gophna, Ram (2008) *From Kerem Matari to Ben Yehouda Strasse: The Story of the Tel Aviv Suburb of Tel Nordau 1922–1932*. Tel Aviv: Gofna [in Hebrew].

Granovetter, Mark S. (1973) "The Strength of Weak Ties." *American Journal of Sociology* 78(6), 1360–1380.

Granovetter, Mark, and McGuire, Patrick (1998) "The Making of an Industry: Electricity in the United States." In: M. Callon (ed.) *The Laws of the Markets*. Oxford: Blackwell, 147–173.

Gross, Nachum (1979) "Then and Now—Comments on the Industrial Census of Palestine 1928." *Economic Quarterly* 26, 11–16 [in Hebrew].

——— (1987) "Absorbing Academic Immigrants in the 1920's and Hydrological Research." *HaZionut* [*Zionism*] 12, 127–140 [in Hebrew].

Gross, Nachum, and Metzer, Jacob (1993) "Palestine in World War Two: Some Economic Aspects." In: G. T. Mills and H. Rockoff (eds.) *The Sinews of War*. Iowa City: University of Iowa Press, 59–82.

Halpern, Ben (1987) *A Clash of Heroes: Brandeis, Weizmann, and American Zionism*. Oxford: Oxford University Press.

Henare, Amiria, Holbraad, Martin, and Wastell, Sari (2007) *Thinking through Things: Theorizing Artefacts Ethnographically*. London: Routledge.

Himadeh, Said B. (ed.) (1938) *Economic Organization of Palestine*. Beirut: American University of Beirut Press.

Hope-Simpson, John (1930) "Report on Immigration, Land Settlement and Development." http://www.zionism-israel.com/Palestine_Hope_Simpson_Report.htm.

Horowitz, Dan, and Lissak, Moshe (1978) *Origins of the Israeli Polity: Palestine under the Mandate*. Chicago: University of Chicago Press.

Hughes, P. Thomas (1983) *Networks of Power: Electrification in Western Society 1880–1930*. Baltimore: Johns Hopkins University Press.

——— (1986) "The Seamless Web: Technology, Science, Etcetera, Etcetera." *Social Studies of Science* 16(2), 281–292.

Huneidi, Sahar (2001) *A Broken Trust: Herbert Samuel, Zionism, and the Palestinians*. London: I. B. Tauris.

Ikein, A., Augustine, Alamieyeseigha, Diepreye, S. P., and Azaiki, Steve (eds.) (2008) *Oil, Democracy, and the Promise of True Federalism in Nigeria*. Lanham, MD: University Press of America.

Jenkins, Keith, and Munslow, Alun (2004) "Walter Benjamin: The Arcades Project." In: S. Jenkins and A. Munslow (eds.) *The Nature of History Reader*. London: Routledge, 135–141.

Kamen, Charles S. (1991) *Little Common Ground: Arab Agriculture and Jewish Settlement in Palestine, 1920–1948*. Pittsburgh: University of Pittsburgh Press.

Karlinsky, Nahum (2000) *Citrus Blossoms: Jewish Entrepreneurship in Palestine 1890–1939*. Jerusalem: Magnes Press [in Hebrew].

Khalaf, Issa (1991) *Politics in Palestine: Arab Factionalism and Social Disintegration, 1939–1948*. Albany: State University of New York Press.

Khalidi, Rashid (2006) *The Iron Cage: The Story of the Palestinian Struggle for Statehood*. Boston: Beacon Press.

——— (2007) "The Palestinians and 1948: The Underlying Causes of Failure." In: E. L. Rogan and A. Shlaim (eds.) *The War for Palestine: Rewriting the History of 1948* (2nd ed.). Cambridge: Cambridge University Press, 12–36.

Kassem, Fatma (2011) *Palestinian Women: Narrative Histories and Gendered Memory*. London: Zed Books.

Kimmerling, Baruch (1983) *Zionism and Economy*. Cambridge, MA: Schenkman.

——— (2004) *Immigrants, Settlers, Natives*. Tel Aviv: Am Oved [in Hebrew].

Kimmerling, Baruch, and Migdal, Joel (2003) *The Palestinian People: A History*. Cambridge, MA: Harvard University Press.

Kline, R. Ronald (2000) *Consumers in the Country: Technology and Social Change in Rural America*. Baltimore: Johns Hopkins University Press.

Latour, Bruno (1987) *Science in Action: How to Follow Scientists and Engineers through Society*. Cambridge, MA: Harvard University Press.

——— (2005) *Reassembling the Social: An Introduction to Actor-Network-Theory.* Oxford: Oxford University Press.

Law, John (1992) "Notes on the Theory of the Actor-Network: Ordering, Strategy and Heterogeneity." *Systems Practice* 5, 379–393.

——— (2008) "Actor-Network Theory and Material Semiotics." In: B. S. Turner (ed.) *The New Blackwell Companion to Social Theory* (3rd ed.). Chichester, UK: Blackwell, 141–158.

Law, John, and Urry, John (2004) "Enacting the Social." *Economy and Society* 33(3), 390–410.

Lentin, Ronit (2008) "The Contested Memory of Dispossession: Commemorizing the Palestinian Nakba in Israel." In: R. Lentin (ed.) *Thinking Palestine.* London: Zed Books, 206–220.

Levi, Ron, and Valverde, Mariana (2008) "Studying Law by Association: Bruno Latour Goes to the Conseil d'État." *Law and Social Inquiry* 33(3), 805–825.

LeVine, Mark (2005) *Overthrowing Geography: Jaffa, Tel Aviv, and the Struggle for Palestine, 1880–1948.* Berkeley: University of California Press.

——— (2007) "Modernity and Its Mirror: Three Views of Jewish-Palestinian Interaction in Jaffa and Tel Aviv." In: S. Sufian and M. LeVine (eds.) *Reapproaching Borders: New Perspectives on the Study of Israel-Palestine.* Lanham, MD: Rowman & Littlefield, 287–306.

Lissak, Moshe (1989) "Immigration and the Construction of Jewish Society." In: M. Lissak, A. Shapira, and G. Cohen (eds.) *The History of the Jewish Yishuv in Eretz Yisrael since the First Aliya: The British Mandatory Period.* Jerusalem: Israeli Academy of Sciences, 345–377 [in Hebrew].

Lockman, Zachary (1995) "Exclusion and Solidarity: Labor Zionism and Arab Workers in Palestine 1897–1929." In: G. Prakash (ed.) *After Colonialism.* Princeton, NJ: Princeton University Press, 211–240.

MacKenzie, Donald, Muniesa, Fabian, and Siu, Lucia (2008) *Do Economists Make Markets?* Princeton, NJ: Princeton University Press.

Malouf, David (1993) *Remembering Babylon.* New York: Vintage Books.

Masalha, Nur (2005) *Catastrophe Remembered: Palestine, Israel and the Internal Refugees.* London: Zed Books.

Melbin, Murray (1978) "Night as Frontier." *American Sociological Review* 43(1), 3–22.

Metzer, Jacob (1982) "Technology, Labor and Growth in a Dual Economy's Traditional Sector: Mandatory Palestine 1921–1936." In: L. Jorberg and N. Rosenberg (eds.) *Technical Change, Employment and Investment.* Lund, Sweden: Lund University, Press, 159–171.

——— (1998) *The Divided Economy of Mandatory Palestine.* Cambridge: Cambridge University Press.

——— (2003) "The Economy of Mandatory Palestine: An Overview of the Development of Research." In: A. Bareli and N. Karlinski (eds.) *Economy and Society in the Mandatory Period.* Beer Sheva: Ben-Gurion University Press, 7–57 [in Hebrew].

Metzer, Jacob, and Kaplan, Oded (1985) "Jointly but Severally: Arab-Jewish Dualism and Economic Growth in Mandatory Palestine." *Journal of Economic History* 45, 327–345.

——— (1990) *The Jewish and Arabic Economies in Mandatory Palestine: Product, Employment and Growth.* Jerusalem: Mossad Bialik [in Hebrew].

Meyer, John W., and Rowan, Brian (1977) "Institutionalized Organizations: Formal Structure as Myth and Ceremony." *American Journal of Sociology* 83(2), 340–363.

Mitchell, Timothy (2002) *Rule of Experts: Egypt, Techno-Politics, Modernity.* Berkeley: University of California Press.

Moberg, Marc (1996) "Crown Colony as Banana Republic: The United Fruit Company in British Honduras, 1900–1920." *Journal of Latin American Studies* 28, 357–381.

Mol, Annemarie (2008) "I Eat an Apple: On Theorizing Subjectivities." *Subjectivity* 22, 28–37.

Mol, Annemarie, and Law, John (1994) "Regions, Networks and Fluids: Anemia and Social Topology." *Social Studies of Science* 24, 641–671.

Morris, Benny (1998) *The Birth of the Palestinian Refugee Problem, 1947–1949.* Cambridge: Cambridge University Press.

——— (2004) *The Birth of the Palestinian Refugee Problem Revisited.* Cambridge: Cambridge University Press.

Moser, Ingunn, and Law, John (1999) "Good Passages, Bad Passages." In: J. Law and J. Hassard (eds.) *Actor Network Theory and After.* Oxford: Blackwell, 196–219.

Murakami, Haruki (2011) *1Q84.* London: Harville Secker.

Nadan, Amos (2006) *The Palestinian Peasant Economy under the Mandate: A Story of Colonial Bungling.* Cambridge, MA: Harvard University Press.

Naor, Mordecai (2003) *The Gift of Electricity.* Jerusalem: Yad Yizhak Ben-Zvi [in Hebrew].

Nye, David (1990) *Electrifying America: Social Meanings of a New Technology.* Cambridge, MA: MIT Press.

Otter, Chris (2008) *The Victorian Eye: A Political History of Light and Vision in Britain 1800–1910.* Chicago: University of Chicago Press.

Passoth, Jan-Hendrik, and Rowland, J. Nicholas (2010) "Actor-Network State: Integrating Actor-Network Theory and State Theory." *International Sociology* 25(6), 818–841.

Platt, Harold L. (1991) *The Electric City: Energy and the Growth of the Chicago Area, 1880–1930.* Chicago: University of Chicago Press.

Ram, Hanna (1996) *The Jewish Community in Jaffa: From Sephardic Community to Zionist Center.* Jerusalem: Carmel [in Hebrew].

Rao, Y. Srinivasa (2010) "Electricity, Politics and Regional Economic Imbalance in Madras Presidency, 1900–1947." *Economic and Political Weekly* 45(23), 59–66.

Rao, Y. Srinivasa, and Lourdusamy, John (2010) "Colonialism and the Development of Electricity: The Case of Madras Presidency, 1900–1947." *Science, Technology and Society* 15(1), 27–54.

Reguer, Sara (1995) "Rutenberg and the Jordan River: A Revolution in Hydro-Electricity." *Middle Eastern Studies* 31(4), 691–729.
Reichman, Shalom, Katz, Yossi, and Paz, Yair (1997) "The Absorptive Capacity of Palestine, 1882–1948." *Middle Eastern Studies* 33(2), 338–361.
Reuveny, Jacob (1993) *The Administration of Palestine under the British Mandate 1920–1948: An Institutional Analysis.* Ramat Gan: Bar Ilan University Press [in Hebrew].
Roberts, Nicholas E. (2011) "Re-Remembering the Mandate: Historiographical Debates and Revisionist History in the Study of British Palestine." *History Compass* 9(3), 215–230.
Rokach, Yitzchak (1970) *Orchard Stories.* Ramat Gan: Masada [in Hebrew].
Rouyer, Awyn R. (1996) "Zionism and Water: Influences on Israel's Future Water Policy during the Pre-State Period." *Arab Studies Quarterly* 18(4), 25–48.
Sa'di, Ahmad, and Abu-Lughod, Lila (2007) *Nakba: Palestine, 1948, and the Claims of Memory.* New York: Columbia University Press.
Samuel, Edwin (1957) *British Traditions in the Administration of Israel.* London: Valentine.
Schivelbusch, Wolfgang (1986) *The Railway Journey: The Industrialization of Time and Space in the Nineteenth Century.* Berkeley: University of California Press.
——— (1995) *Disenchanted Night: The Industrialization of Light in the Nineteenth Century.* Berkeley: University of California Press.
Segev, Tom (1999) *Palestine under the British.* Jerusalem: Keter [in Hebrew].
——— (2000) *One Palestine, Complete: Jews and Arabs under the British Mandate.* New York: Henry Holt.
Shaffer, Simon (1997) "Metrology, Metrication, and Victorian Values." In: B. V. Lightman (ed.) *Victorian Science in Context.* Chicago: University of Chicago Press, 438–474.
Shafir, Gershon (1993) "Land, Labor, and Population in Zionist Colonization: General and Specific Aspects." In: U. Ram (ed.) *Israeli Society: Critical Perspectives.* Tel Aviv: Breirot Publishers, 104–119 [in Hebrew].
Shalev, Michael (1989) "Jewish Organized Labor and the Palestinians: A Study of State/Society Relations in Israel." In: B. Kimmerling (ed.) *The Israeli State and Society: Boundaries and Frontiers.* New York: State University of New York Press, 93–133.
Shaltiel, Eli (1990) *Pinhas Rutenberg: 1879–1942, Life and Times.* Tel Aviv: Am Oved.
Shamir, Ronen (2000) *The Colonies of Law: Colonialism, Zionism, and Law in Early Mandate Palestine.* Cambridge: Cambridge University Press.
——— (2001) "Jewish Bourgeoisie in Colonial Palestine." *Israeli Sociology* 1, 133–148 [in Hebrew].
Shamir, Ronen, and Hacker, Dafna (2001) "Colonialism's Civilizing Mission: The Case of the Indan Hemp Drugs Commission." *Law and Social Inquiry* 26(2), 435–461.
Shapira, Anita (1989) "Political History of the Yishuv 1918–1939." In: M. Lissak, A. Shapira, and G. Cohen (eds.) *The History of the Jewish Yishuv in Eretz Yisrael since the First Aliya: The British Mandatory Period.* Jerusalem: Israeli Academy of Sciences, 1–171 [in Hebrew].

——— (1992) *Land and Power: The Zionist Resort to Force 1881–1948*. New York: Oxford University Press.

Shapiro, Yonatan (1973) "The Debate between Chaim Weizmann and Louis Brandeis 1919–1921." In: D. Carpi (ed.) *Zionism: Studies in the History of the Zionist Movement and of the Jews in Palestine*. Tel Aviv: Tel Aviv University Press, vol. 3, 258–272.

Shenkar, Arie (1963) *Be'emuna U'be'maas*. Tel Aviv: privately published [in Hebrew].

Shoham, Hizky (2012) "Tel Aviv's Foundation Myth: A Constructive Perspective." In: M. Azaryahu and I. Troen (eds.) *Tel Aviv, the First Century: Visions, Designs, Actualities*. Bloomington: Indiana University Press, 34–59.

Smith, J. Barbara (1993) *The Roots of Separatism in Palestine: British Economic Policy 1920–1929*. London: I. B. Tauris.

Stalder, Felix (2002) "Failures and Successes: Notes on the Development of Electronic Cash." *The Information Society* 18(3), 209–219.

——— (2005) *Open Cultures and the Nature of Networks*. Sarajevo: Futura Publishers.

——— (2006) *Manuel Castells: The Theory of the Network Society*. Cambridge: Polity Press.

Star, Susan L., and Griesemer, James R. (1989) "Institutional Ecology, Translations and Boundary Objects: Amateurs and Professionals in Berkeley's Museum of Vertebrate Zoology, 1907–39." *Social Studies of Science* 19(3), 387–420.

Sternhell, Zeev (1995) *Nation-Building or a New Society?* Tel Aviv: Am Oved [in Hebrew].

Sufian, Sandra M. (2007) *Healing the Land and the Nation: Malaria and the Zionist Project in Palestine, 1920–1947*. Chicago: University of Chicago Press.

Sunderland, David (2004). *Managing the British Empire: The Crown Agents, 1833–1914*. Woodbridge, UK: Boydell & Brewer.

Sussman, Zvi (1973) "The Determinants of Wages for Unskilled Labor in the Advanced Sector of the Dual Economy of Mandatory Palestine." *Economic Development and Cultural Change* 22(1), 95–113.

Tinker, Hugh (1968 [1954]) *The Foundations of Local Self-Government in India, Pakistan, and Burma*. London: Pall Mall Press.

Trottier, Julie (1999) *Hydropolitics in the West Bank and Gaza Strip*. Jerusalem: PASSIA (Palestinian Academic Society for the Study of International Affairs).

Valderrama Pineda, Andrés (2010) "How Do We Co-Produce Urban Transport Systems and the City? The Case of Transmilenio and Bogotá." In: I. Farías and T. Bender (eds.) *Urban Assemblages: How Actor-Network Theory Changes Urban Studies*. Abingdon, UK: Routledge, 123–138.

Valderrama Pineda, Andres, and Jorgensen, Ulrik (2008) "Urban Transport Systems in Bogota and Copenhagen: An Approach from STS." *Built Environment* 34(2), 200–217.

Veeser, Cyrus (Forthcoming) "A Forgotten Instrument of Global Capitalism? International Concessions, 1870–1930." *International History Review*.

Waterbury, John (1979) *Hydropolitics of the Nile Valley*. Syracuse, NY: Syracuse University Press.

Weiss, Dana, and Shamir, Ronen (2011) "Corporate Accountability to Human Rights: The Case of the Gaza Strip." *Harvard Human Rights Journal* 24, 155–183.

Weizmann, Chaim (1949) *Trial and Error*. London: Hamish Hamilton.

Whitehead, Mark (2008) "Cold Monsters and Ecological Leviathans: Reflections on the Relationships between States and the Environment." *Geography Compass* 2(2), 414–432.

Wilhite, Harold (2005) "Why Energy Needs Anthropology." *Anthropology Today* 21(3), 1–2.

Wilhite, Harold, Nakagami, Hidetoshi, Masuda, Takashi, Yamaga, Yukiko, and Haneda, Hiroshi (1996) "A Cross-Cultural Analysis of Household Energy Use Behavior in Japan and Norway." *Energy Policy* 24(9), 795–803.

Winther, Tanja (2008a) "Empowering Women through Electrification: Experiences from Rural Zanzibar." *Energia News* 11(1), 8–10.

——— (2008b) *The Impact of Electricity: Development, Desires and Dilemmas*. Oxford: Berhahn Books.

Yaari-Poleskin, Jacob (1939) *Pinhas Rutenberg*. Tel Aviv: Ha-Tekufah [in Hebrew].

Yahav, Dan (2009) *Electricity Flow in Her Veins*. Tel Aviv: Shaked [in Hebrew].

Ziman, Yehoushoua (1929) *Jewish Economy in Numbers*. Tel Aviv: Davar [in Hebrew].

Zu'bi, Nahla (1984) "The Development of Capitalism in Palestine: The Expropriation of the Palestinian Direct Producers." *Journal of Palestine Studies* 13(4), 88–109.

Zureik, Elia T. (1979) *The Palestinians in Israel: A Study in Internal Colonialism*. London: Routledge & Kegan Paul.

Index

Acre, 22, 74, 133
actor networks, 9–13, 94, 102, 118, 125–129, 149, 152–156, 165n7, 173nn39–40, 177n37, 181n20 (*see also* networks); attachment to, 2, 13, 23, 27, 49, 56, 60, 64, 70, 80–81, 99–100, 111, 115, 121, 124, 132–133, 145–146, 148, 150–151, 157n1, 174n53, 183n43; boundary object, 94, 118, 142–143, 173n40; context and, 7–8, 27–34, 37, 51, 152, 161–162n10, 181n20; durability, 155; flattened topography and, 7, 82, 85, 152; heterogeneity of, 7–8, 12, 45, 51, 102, 152–155; identity and, 9, 70, 143; as immutable mobiles, 10, 157–158n4; as intermediaries, 5–6, 18, 118; as mediators, 5–6, 111, 118; nonhumans and, 10, 12, 102, 149, 151; regions and, 2–3, 6, 16, 48–49, 52, 54–56, 58, 61, 174n54; as single-point actors, 152–156; sociology of associations and, 8, 10; subjectivity and, 9, 99; theory, 9–13, 129, 149, 152–156, 181n21
agriculture, 8–9; Arab, 13, 145, 148; electric grid and, 23, 119, 124, 145; electrification of, 111; Jewish, 19, 138, 144, 147–148, 155; legislation, 183n40; products, 136, 180n15; science, 124, 145, 155. *See also* citrus farming; irrigation; water, wells
Amery, Leo, 28, 172n24, 181n27
Anglo-Palestine Bank, 26–27, 161n4, 178n39
appliances, electric, 1, 98, 101, 115, 163n30
Arab economy, 18–19, 134–137. *See also* agriculture, Arab; economy; industry, Arab
Arab-Jewish relations, 10, 15, 18, 22–23, 28, 30, 43–44, 77, 132–133, 161–162n10, 166n21; cooperation, 20, 29, 48; divide, 17, 21, 58, 65, 70, 78, 148–150; relational approach, 20
Arab-Palestinian: farmers, 52, 76, 124, 144, 146–148; labor, 65, 147; landowners, 26–29, 32, 49, 55, 61–62; leadership, 15, 27, 161n4; nationalism, 5, 27, 76, 156; newspapers, 62, 64–65, 73, 167n33, 167n36, 169n58; population, 16–17, 21–22, 26–27, 160n20, 170n4; settlements, 22, 71–72, 74, 76, 118, 142, 169n60
Arab-Palestinians and electricity: attachment to electric grid, 71–72, 74–77, 101, 118, 132, 146–148, 150, 155, 183n43; consumers, 64–65, 167n35; opposition to concessions, 15, 21, 26–27, 46, 72, 161n6, 167n35; resistance to, 27–28, 32–33, 40, 62–63, 70, 72–74, 156, 160–161n3, 167n33, 167n36, 169n60
Arab revolt of 1936, 17. *See also* separatism
Auja River (Yarkon), 24–26, 28–30, 33, 35, 38, 62, 170n4; concession,

Auja River (Yarkon) (*continued*) 15–16, 24, 29–30, 35–36, 38, 42, 48, 81, 115, 118, 130, 137–138, 156, 158n8, 161–162n10, 165n6, 181n22 (*see also* concessions); hydroelectric plan, 15, 24–25, 27–30, 32–33, 34–39, 45–46, 61, 163–164n37 (*see also* hydroelectric stations); irrigation, 141–142, 182n31, 182n34, 184n45 (*see also* irrigation); water rights, 138–141, 148, 181n27 (*see also* Mavromatis, Euripides)

Baharav, Yekutiel, 52
Balfour, Arthur James, 13
Balfour Declaration, 13, 21, 161–162n10
Barker, Francis, 109
Be'er Sheba, 104
Beitar, Omar, 166n27, 167n34
Bentwich, Norman, 89, 118, 169n51, 172nn21–24
Berkenheim, Alexander, 159n15
Bernstein, Deborah, 134
Bethlehem, 76, 169–170n61
Britain. *See* Great Britain
British government, 11, 29, 42, 107–108, 112, 114, 120, 154, 159n10, 161–162n10; Colonial Office, 11, 15–16, 28, 31–35, 42, 44–46, 73, 88–89, 106–111, 113, 119–121, 138–141; Foreign Office, 95, 175n4, 175n6
Burchells (law offices), 88, 162n21
Burma, 166n17
Buzelin, Helene, 12

Callon, Michel, 134, 181n19
capitalism, 123, 148, 151, 158n9, 160n21
Churchill, Winston, 15, 28, 36, 42–45, 111, 159n12, 163n29, 176n19, 176n23
citrus farming, 7, 11, 61, 76, 85, 118, 124, 137, 140, 142–147, 181n22, 182n31, 183nn41–42. *See also* agriculture; irrigation
colonialism, 14, 17, 27, 134, 151, 156
Colonial Office. *See* British government, Colonial Office
concessions, 158n9; colonial practice of, 14–15, 18, 26–27, 159n10 (*see also* Palestine, government of); electricity, 14, 18, 55, 94–95, 98 (*see also* Electricity Ordinance); railways, 108. *See also* Arab-Palestinians and electricity, opposition to concessions; Auja River, concession; Jordan River, concession; Mavromatis, Euripides, concession
Council of the Township of Tel Aviv, 37–39, 60–61, 163n30. *See also* Dizengoff, Meir
current meter, 93–103, 117, 123, 129, 155

diesel fuel, 1, 31, 112; engines, 25, 36, 37–39, 45, 134, 138, 163–164n37, 164n45; powerhouse, 28–29, 38–40, 45, 47, 61, 73, 133, 139, 146, 163–164n37
distribution network, electric, 2, 23, 36, 55, 60–61, 83, 94, 96–99, 103, 117, 126, 142, 155
Dizengoff, Meir, 37–39, 49–50, 60, 67, 69, 163n31, 167n32, 168n44
dual-society paradigm, 19–20, 160n22

East Jerusalem Electric Company, 169–170n61
economics, 134, 181n19
economists, 19, 74, 78, 100, 101, 132, 134–136
economy, 8, 14, 22, 40, 55, 69, 74, 100, 114, 121–122, 134–135, 143; British policy, 18–19, 132 (*see also* Hope-Simpson Report); development, 15, 22, 26, 43, 111, 150; divided, 18–19, 132–133, 150; dual, 18, 155. *See also* Arab economy; Jewish economy
Egypt, 7, 13, 67–68, 95, 104, 108, 169–170n61
electric current, 1–2, 6, 22, 37, 46, 51, 60, 62, 74, 91–92, 106, 109, 115–116, 133–134, 157n1, 157n3, 165n4
electricians, 2, 6, 83, 90, 97, 155
electricity, 1–8, 11, 22, 24, 31; access to, 86, 124; appliances, 1, 98, 101, 115,

163n30; consumers, 98–103; demand, 38, 45, 120–121, 128, 132; direction of, 104; irrigation, 138–139, 144, 146, 148 (*see also* irrigation); law and, 6, 81–84, 88–89, 94–95, 102, 117–118; as luxury, 72, 74; nationalism and, 70, 72, 74–76, 169–170n61 (*see also* Arab-Palestinians and electricity; ethno-national communities, electricity and); public safety and, 90; railways, 106, 109, 111–115 (*see also* railways, electrification of); rates, 43, 93, 95, 166–167n30; regime, 80–81, 83, 93, 102, 117–118; supply, 24–25, 27, 34, 55, 59–60, 68–69, 71, 89, 106, 137 (*see also* industry, electric supply); units of measurement, 93 (*see also* current meter)
Electricity Ordinance, 88–90, 94–95, 169n51; draft legislation, 70, 88–90, 95, 98, 169n52, 172n22, 172n27
electric meter. *See* current meter
electrification, 1–2, 5–6, 8, 11, 20–21, 23, 74, 82–83, 86, 151, 154–155; in Africa, 2, 5, 14, 20, 72, 158n5, 159n10; in Canada, 75; in colonies, 4; in Europe, 3, 55, 68, 83; in India, 5, 54; rural, 8, 71, 74–76, 100–101, 158n5; in the United States, 3, 68; urban (*see* urban space, electrification of); Zionism and, 3, 18, 26, 60 (*see also* separatism). *See also* railways, electrification of; tramways
El-Husseini, Moussa Kazin, 161n6
engineers, electrical, 2, 28, 34–36, 39, 51, 60, 84–85, 88, 109–113, 115–117, 164–165n46
English Electric Company, 41, 109
ethno-national communities, 5, 7, 9, 20–21, 58–59, 65; distinctions, 9, 21–22, 27, 56, 60, 102, 147, 150, 154; divide, 21–22, 137, 150, 155; electricity and, 70, 148, 183n43; statistics of, 132, 134; tensions, 26, 166n21
Europe, 2–4, 38, 68, 84, 102, 106, 109, 121–123, 157n2, 165n4, 175n56
experts, 11, 28, 34, 42, 60, 106–107, 112, 116–117, 138, 171n8, 176n7, 176n23, 181n26, 182–183n38, 183n40

Faris, Basim, 74–75, 92, 100–102, 132, 173n36

Gandhi, Mahatma, 54
Gaza, 104; strip, 169–170n61
General Electric Corporation, 40, 68
Germany, 3, 28, 30, 39–40, 43–45, 50–51, 54, 67, 111, 143, 163n35, 168n45; Berlin, 28–29, 37–42, 45–47, 168n45; electrical engineers, 28, 39; electrical equipment, 43, 120; firms, 38–41, 45, 47–48, 67, 84, 111, 168n45, 171n8, 177n26; imports, 43–44, 50–51, 163n35, 164n42; machinery, 28, 30, 41–42, 115
Granovetter, Mark, 4, 54, 151
Great Britain, 3, 13, 40, 42–45, 107, 117, 122, 139, 159n14, 175n4; army, 13, 104, 108, 175n1; electrical equipment, 42–45, 50; firms, 8, 26, 41, 45, 109, 119–120, 134, 164–165nn45–46, 175n4
grid, electric, 1–2, 4–8, 54–56; access to, 22, 38, 48, 78, 80, 90–92; actor-network theory and, 13, 99, 155 (*see also* actor networks); construction, 47, 60, 71, 75; ethno-national distinctions and, 9, 49, 61–62, 65, 76–77 (*see also* ethno-national communities); expansion, 8, 51, 54–55, 63–64, 76, 86, 90–91, 98, 101–102, 119, 122–124, 156; law and, 82–83, 117–118; as social formation, 51, 80–81, 102, 129–130; statistics, 88, 123; stratification and, 78, 99–103; as technical network, 11, 155. *See also* current meter; lights, electric; poles, electric; wires, electric
Gross, Nachum, 143
group formation, 3, 7, 20, 22, 100

Haifa, 22, 70–73, 107, 124, 133–134, 145–146, 165n13, 168n49, 169n57, 173n36, 180nn15–16; municipality, 73; powerhouse, 73, 115–116, 119, 133–134, 169n56 (*see also* powerhouse)
Hedjaz Railway, 104, 107–108, 162n19, 175n1, 175nn3–4. *See also* railways
High Commissioner of Palestine, 15–16, 31, 89, 95–96, 106, 108–111; Plumer, Herbert, 28. *See also* Samuel, Herbert
historical sociology, 149
Holmes, Robert, 110, 175n4
Hoofien, Siegfried, 161n4
Hope-Simpson Report, 142–144, 146–147, 182–183n38, 183n40, 183–184n44
Hughes, Thomas, 3–4, 11, 40, 54–55, 104–106, 165n14, 166n16, 168n45, 175n56
hydroelectric stations, 1, 21, 25, 28, 31–33, 54; turbines, 41. *See also* Auja River; Jordan River
hydrology, 140, 144; hydrologists, 138, 143, 146
hydro-politics, 137–138, 143–144, 181n21

Ibn-Saud (King), 175n2, 175n4
immigration. *See* Jewish immigration
import duties, 126, 128
India, 5, 54–55, 113, 135, 157n1, 172n28
industrialization, 4, 72, 84, 106, 134
industrial power, 129–131, 133, 135, 137–138, 179n51 (chap. 4), 179n8 (chap. 5)
industry: Arab, 137; electric, 4, 68, 74; electric supply and, 22, 37–38, 61, 68, 71, 81, 106, 111, 118–119, 121, 124, 126–128, 131–132, 133–134, 145; Jewish, 18–19, 121, 123–124, 128, 133, 145; Palestine statistics, 132–136
infrastructure, 1, 4, 6, 14, 22, 27, 56, 77, 83, 104, 107, 135, 142, 150, 158n6, 165n7, 166n26, 171–172n17
Institute of Electrical Engineers, 109, 162n18
international law, 138, 159n14; Permanent Court of International Justice, Hague, 16, 139, 159n14. *See also* law and legislation
Internet, 1, 80–81, 166n26
invention, 4, 100, 106
irrigation, 1, 7, 33, 118, 130, 139–140, 142, 144–148, 156; electricity and, 34, 37, 51, 81, 124, 129–130, 137–140, 142, 146, 148, 173n34, 173n38, 181n27, 182n31, 182–183n38, 183n42, 184n45. *See also* agriculture; water
Israel Electric Corporation, 11, 169–170n61

Jaffa, 11, 19, 22; Ajami quarter, 48, 52–53, 64; Chamber of Commerce, 63; committee of merchants, 50; Customs House, 115; King George Boulevard, 52, 63; Manshie quarter, 58; mayor of, 37, 59, 166n27, 167–168n39; municipality, 29, 31–33, 38, 46, 55–56, 59–65, 71, 152, 161n4, 166n20, 167n35, 168n41, 169n60; municipal planning committee, 51, 64, 166n20; population, 24, 65, 72, 122, 140, 167n35; port, 47, 68, 107, 115; powerhouse, 24–30, 39, 45, 61, 106, 138–141 (*see also* powerhouse); Tel Aviv and, 24, 27, 34, 37–39, 49, 55–56, 58–62, 64–65, 91, 122–123, 166n21, 171n16
Jaffa district, 11, 15, 24, 34, 39, 45, 48, 53, 55–56, 65, 67–68, 72, 81, 102–106, 114–115, 133, 138–139, 141, 145, 167n36
Jaffa Electric Company, 24–25, 47, 65, 95; acquisition of land, 53, 55; annual reports, 51, 71, 88, 92, 101, 122–123; equipment, 40, 45–47, 164–165n46; Jaffa municipality and, 46, 59, 61–64, 72, 161n4; legal issues, 52, 81–82, 126–127, 138–142; Palestine Electric Company and, 11, 16, 131, 178n47 (*see also* Palestine Electric Company); Tel Aviv municipality, 62, 83–84, 123; tramway, 67–70

Jaffa–Jerusalem Railway, 43, 56, 67, 104, 107–110, 175n5, 176n7; electrification of, 43, 108, 110–111, 113–115, 119–120, 176n9, 177n25. *See also* railways
Jerusalem, 11, 13–16, 28, 33–37, 40, 42, 45–46, 56, 74, 76, 89, 106, 108–109, 126; water supply, 15, 138–141, 182nn34–35
Jerusalem District Electric Company, 169–170n61
Jewish Agency, 133, 136, 183n40, 183–184n44
Jewish-Arab relations. *See* Arab-Jewish relations
Jewish Colonial Trust, 26, 161n4
Jewish Colonization Association, 26, 183–184n44
Jewish economy, 19, 126, 134–138, 145, 147, 155–156; manufacturing and, 136; private capital, 19, 121–124, 146; private farming, 144–145, 181n22. *See also* agriculture, Jewish; economy; industry, Jewish
Jewish immigration, 18–19, 85, 102, 122–124, 132, 143–145, 166n21, 171n16, 178n48
Jewish Industrial Association, 136
Jewish national home, 14, 26, 42, 136, 158n7, 161n6, 161–162n10 (*see also* Zionism)
Jewish population, 17–18, 27, 72, 121, 136–137, 150, 160n20
Jewish settlements, 71–72, 76, 114, 118, 124, 132, 145–148, 169–170n61, 178n38
Johnson-Lodge Immigration Act, 122
Jordan River, 15, 21, 38, 73, 106, 115, 184n45; concession, 15–16, 73, 95, 113, 161–162n10, 174n52, 177n28, 180n11 (*see also* concessions); hydroelectric station, 21, 73, 106, 109, 111, 115, 119–120, 134, 176n12, 177n28 (*see also* hydroelectric stations)

Keren Hayesod, 136
Keynes, John Maynard, 135
Kfar Saba, 183n42
Kisch, Frederick (Colonel), 126
Kline, Ronald, 8, 70–71, 74–75, 101, 106

Latour, Bruno, 5, 8, 12, 100, 152, 157–158n4
Law, John, 19, 48–49, 151, 154
law and legislation, 55, 73, 88–89, 127–128, 144, 183n40. *See also* electricity, law and; regulation; techno-legal regime
lawyers, 6, 15, 28, 30, 32, 35, 36, 39, 47, 117, 141, 169n56
League of Nations, 13, 26
lights, electric, 2, 23–24, 37–38, 63, 80–84, 92, 101, 114, 130, 148, 167–168n39 (*see also* streetlights); consumers of, 122–123; nationalism and, 27; private, 81, 92, 101, 114, 118, 122–123, 133 (*see also* private and public spheres); public, 34, 51, 81, 86, 93, 114, 123, 130, 173n32; security and, 84–85; statistics of, 129, 131–132, 156
Lydda, 71, 74, 104, 108, 118

Mandate of Palestine, 13, 18, 20, 26, 132, 179n61; government of, 32, 88 (*see also* Palestine, government of); state of, 22, 83, 132, 136, 138, 143, 148
maps, 6, 11, 50–51, 54, 56, 58–59, 61, 64, 71, 86, 87, 140, 149, 155, 183n43
markets, 9, 134; concessions and, 14, 26; for electricity, 40, 71–72, 106, 132; labor, 18, 160n22; and society dichotomy, 135; and state dichotomy, 8, 113, 119, 124
Mavromatis, Euripides, 15, 141, 160–161n3; concession, 15–16, 76, 138–140, 159n14, 169–170n61, 182n34. *See also* concessions
Melchet, Lord. *See* Mond, Alfred
Metzer, Jacob, 133–137, 180n14
micro-macro dichotomy, 151–152, 154
Mitchell, Timothy, 7, 134–135

mobile phone, 80
Mond, Alfred, 26, 120–121, 127, 134, 163n28, 178n45
Moslem-Christian Association, 62, 166n27

Nablus, 76
Nadan, Amos, 18
Nassatisin, Michael, 26
nationalism, 74, 76, 151, 156, 170n4. *See also* Arab-Palestinian, nationalism; Zionism
Nazareth, 22, 75
Nes-Ziona, 118
networks, 49, 129–130, 154–155; actor-network theory and, 12–13, 153 (*see also* actor networks); communication, 80; electric, 3, 54; interpersonal, 151–152; social, 4, 6, 12, 151–152; technical, 11–12; transportation, 68, 104, 106–107
Nye, David, 7–8, 68, 165n4, 168nn45–46

orange groves. *See* citrus farming
Otter, Chris, 83, 85, 171–172n17
Ottoman Empire, 13, 26; concessions, 15–16, 108, 138–139, 175n5; law, 144; railways, 67, 104, 107–108. *See also* railways

Palestine: British conquest of, 13; census, 1922, 26; partition plan, 17, 159n17; population, 22, 33, 132; West Bank, 76, 169–170n61
Palestine, government of, 70, 75, 88–89, 95, 107–109, 111, 132–133; Auja Concession and, 30–33 (*see also* concessions); civil administration, 14, 108, 134; as consumer of electricity, 22, 112–116; Customs Exemption Ordinance, 126–128; Department of Commerce and Industry, 109–110, 119, 128; Department of Customs, Excise and Trade, 163n35; Department of Health, 170n4; Department of Public Works, 51, 81, 84, 89, 115, 141, 168n49; legal adviser (*see* Bentwich, Norman). *See also* Electricity Ordinance; High Commissioner of Palestine; Samuel, Herbert
Palestine Economic Society, 135
Palestine Electric Company, 11, 16, 26, 92, 95, 131, 139, 142–148, 163n28, 169–170n61, 184n45; annual reports, 129–132; board of directors, 24, 26, 92, 120, 127. *See also* Trade Facilities Act
Palestine Railways, 14, 42, 106–115. *See also* railways
Paris, France, 84
Peel Royal Commission of Inquiry, 17, 160n22
Permanent Court of International Justice, Hague. *See* international law
Petah-Tikva, 37, 67, 124, 139, 142, 178n38, 181n22, 181n27, 183n42
Plender, William, 178n44
Polanyi, Karl, 135
poles, electric, 2, 6, 8, 11, 22, 27, 41, 47–48, 50–53, 55–56, 59–64, 78, 80, 82–83, 86, 88, 98, 102, 155–156, 165n6, 171n8
political economy, 9, 18
political sociology, 13
politics: of calculation, 134–135, 137, 181n19; identity, 5; regional, 5, 54; of separatism, 19, 148 (*see also* separatism); technology and, 7, 30. *See also* hydro-politics
powerhouse, 2, 29, 164n45, 169–170n61. *See also* Auja River, hydroelectric plan; diesel fuel, powerhouse; Haifa, powerhouse; hydroelectric stations; Jaffa, powerhouse; Jordan River, hydroelectric station; Tiberias
Preece, Cardew & Rider, 34–37, 40, 88–89, 110, 113, 115, 162n18, 176n13
private and public spheres, 3, 23, 96–97, 99, 102–103, 117

progress, 8, 63, 73, 90, 100, 169n58
Protocol of Lausanne, 16
public sphere. *See* private and public spheres
pumps, electric. *See* water, pumps, electric

Rafah, 104, 108
railways, 1, 7, 22–23, 104, 106–107, 159n10, 161–162n10; department of, 115; electrification of, 34, 43, 69, 106, 109–115, 119–121, 124, 176n9, 176nn11–12; workshops, 115–116, 119. *See also* Hedjaz Railway; Jaffa–Jerusalem Railway; Palestine Railways; tramways
Ramallah, 76, 169–170n61
Ramat Ha-Sharon, 147
Ramleh, 22, 71, 74, 118, 178n39
Reading, Lord (Rufus Isaacs), 26, 164–165n46
Reguer, Sara, 113, 163n28, 176n12, 176n23
regulation, 81–83, 89–90, 94–95, 98, 125, 143–144. *See also* electricity, law and; law and legislation
Rehovot, 118, 169n53, 183n42
Rishon Le-Zion, 71, 118, 169n53, 178n39, 183n42
Rothschild, Edmond James de, 13, 26
Rutenberg, Abraham, 28
Rutenberg, Pinhas, 14–15, 158n8, 159n11, 159n15; Arabic language, 72, 169n55; Arab nationalism, 26–27, 33, 62, 64, 73, 161n6, 167n34, 167n36 (*see also* Arab-Palestinians and electricity); Auja hydroelectric plan, 25, 34–36; Churchill, Winston, 42–44; Customs Exemption Ordinance, 126–128; Dizengoff, Meir, 38–39, 50; Electricity Ordinance, 89–90, 169n51, 172n27; Jaffa, 31, 33, 161n4, 168nn40–41; railway electrification, 108–113; Sacher, Harry, 28–29; Samuel, Herbert, 108, 115; Weizmann, Chaim, 42, 164n39

Sacher, Harry, 28–30, 159n15
Said, Asim, 166n27, 167n34
Samuel, Herbert, 15, 28, 31, 35, 44, 73, 108–109, 115, 139, 176n23
Sarafand (British Army Headquarters), 53, 71, 74, 108, 116, 177n34, 177n36; electrification of, 116–118
Schivelbusch, Wolfgang, 7, 84, 171n14
science and technology studies, 9–10, 165n7
separatism, 10, 17–19, 21, 23, 160n19, 160n22; electrification and, 150
Shaltiel, Eli, 158n8, 159n11
Shuckburgh, John, 162n19, 172n27
Sierra Leone, 35, 162n22
Smith, Barbara, 18, 44, 167n33, 169n60
Snell, John, 110, 162n18, 163n28, 176n13
Statistical Abstract of Palestine, 136
statistics, 65, 67, 100, 103, 132–134, 135, 146, 163n35, 168n45, 177n37. *See also* ethno-national communities, statistics of; grid, electric, statistics; industry, Palestine statistics; lights, electric, statistics of
Sternhell, Zeev, 19, 158n8
streetlights, electric, 11, 23, 31, 56–57, 60, 63, 72–73, 78, 81, 83–86, 90–91, 123, 129–130, 133, 155, 161n4, 167n35, 168n41, 169n60, 171n14, 173n38, 178n39, 179–180n9; effects of, 92; policing and, 84–85
Sufian, Sandra, 170n4
Syria, 68, 74, 107, 132, 175n4
systems, electric, 3–4, 104

techno-legal regime, 80, 93–94, 102, 117–118, 155
technological determinism, 3
technology transfer, 4
technoscientific state, 143
Tel Aviv, 11, 19, 24, 45–46, 55–56, 58, 68, 101, 122, 133; Manshie quarter, 58; Municipal Department of Light, 84; municipal electrical engineer,

Tel Aviv (*continued*)
84–85; municipality, 11, 31, 60, 123, 166–167n30; population, 24, 122; Tel Nordau, 172–173nn29–31. *See also* Council of the Township of Tel Aviv; Jaffa, Tel Aviv and
telegraph, 4, 11, 28, 80, 173n39
Tiberias, 73, 146, 173n36, 180n16. *See also* powerhouse
Trade Facilities Act, 119–120
Trade Facilities Advisory Committee, 119–121, 178n44
tramways, 65–70, 106, 114, 168n44
Trans-Jordan, 13, 15, 107, 175n4, 178n47
transportation systems, 4, 14–15, 65, 70, 106; urban, 68–69. *See also* railways

United States, 3–4, 54, 68–71, 74–75, 101, 106, 109–110, 121–122, 138, 158n7, 165n4
urban space: electrification of, 4, 7–8, 68, 71, 74–76, 85, 90, 92, 100–101, 106; immigrants and, 123–124; rural and, 8, 71, 75, 101 (*see also* electrification, rural). *See also* private and public spheres

Vernon, Roland Venables, 32, 161n7, 162n24, 177n24, 177n27

water, 1, 75–76; as boundary object, 142–143; immigration policies and, 143–144 (*see also* hydro-politics); institutes, 147; irrigation and, 33, 137–138, 142–147 (*see also* irrigation); pumps, electric, 34, 119, 124, 137–138, 145–146, 148, 155; regime, 143–144, 183n40; rights, 138–141, 144, 148, 155, 181n27; supply, 15–16, 51, 55, 60, 81, 92, 114, 118, 123, 129–131 (*see also* Jerusalem, water supply; Mavromatis, Euripides;); wells, 11, 142–146, 148, 183n43. *See also* Auja River, hydroelectric plan
Weber, Max, 150
Weizmann, Chaim, 42, 160n21, 164n39, 164–165n46
West Bank, 76, 169–170n61
Westinghouse, 68
wires, electric, 1–2, 8, 11, 48, 50–53, 55, 61, 91, 94, 150, 154–155; actor network and, 5–6, 18, 118; copper, 2, 41, 45, 50, 59, 164n42, 165n4; high-tension, 1, 56, 76, 81, 88, 178n39; of Jaffa, 56, 63; laying of, 73, 80, 83, 86, 96, 102; low-tension, 1, 56, 58, 88, 101; as property, 60, 62; of Tel Aviv, 56
World War I, 3, 26, 104, 107–108, 166n16

Yarkon River. *See* Auja River

Zanzibar, 84, 158n5
Zeidner, Yosef, 37–38
Zimbabwe (Rhodesia), 5, 14, 20, 72, 84
Zionism, 17–19, 21, 150, 156, 158n7, 161–162n10, 170n4; Electric Company and, 72, 75, 148, 150, 158n8, 167n32; organizations, 14, 18–19, 21, 26–27, 124, 135–136, 143–144, 159n15, 183n40, 183–184n44
Zionist Executive Committee, 27, 42, 46, 126–128; small industrialists and, 126–127
Zionist organizations, 13, 26, 127, 135, 143, 159n15, 160n21